ISBN-10: 0997326468
ISBN-13: 978-0-9973264-6-8
Third Edition, July 2020

LEARN CODING BASICS IN HOURS
WITH JAVASCRIPT

An Introduction to Computer Programming for People with No Prior Experience

**Written by: Jack C. Stanley and Erik D. Gross,
Co-Founders of The Tech Academy**

TABLE OF CONTENTS

CHAPTER ONE
WELCOME

Welcome to The Tech Academy's *Learn Coding Basics in Hours with JavaScript* book!

Now, we're sure you want to dive right into coding but there's some information we are going to cover first to ensure you are well prepared. In fact, you won't write any code until the 13th chapter of this book.

It may seem like a lot of reading before we get down to business but it's worth it. Let's compare this to another subject: how hard would it be to drive a car if you couldn't define the important terms – such as steering wheel, gas pedal or brake pedal? So, please bear with us and take some time learning some fundamental data that will set you up well for the adventure that lies ahead.

We have attempted to define all technical terms used in this book but if you come across any that you don't understand, look these up in The Tech Academy's *Technology Basics Dictionary* (available for purchase on Amazon).

The Tech Academy is a technology school headquartered in Portland, Oregon, with multiple campuses around the world. We specialize in coding boot camps (intensive training programs aimed at preparing graduates for entry-level technology positions, that can be completed in a few months). We also offer custom training for companies, and have published several books.

While we would love to tell you everything about computers and technology in this book, we are only going to cover the absolutely necessary data. If you are interested in learning about how computers work exactly, we recommend reading our book *You Are Not Stupid – Computers and Tech Simplified*, which can be purchased on Amazon.

Our educational philosophy is to *assume no prior knowledge on the part of the student*. What this means is that our training content, including this book, is written for absolute beginners. We operate on the assumption that you've never written code before and that you don't know basic technical terms – this ensures nothing is over your head and that you aren't lost while reading.

You're about to experience The Tech Academy's approach to education first hand: we are going to start our book by defining the technical words used in the title of our book!

CODING

Coding refers to the action of entering instructions into computers to make them perform specified actions. A computer is a special type of machine.

Machines are devices (equipment with a purpose; tools) made by humans to get work done. They are usually made out of durable materials like wood, plastic and metal. Normally they have some parts that move and some parts that don't; sometimes they have no moving parts at all. They receive some kind of energy that they use to do their work.

One of the traits that makes people different from animals is their ability to create complex machines.

Usually people create machines because there is some work they want to do that the machine could help them with. The help the machine provides could be to get the work done faster, to do the work with less chance of errors, or to do the work nearly continuously, without the need to stop for food or sleep. There are other reasons people make machines, but it usually comes down to getting more work done in a given amount of time with fewer errors.

As time goes on, machines often get improved or changed to make them more effective or to respond to changes in the area of society where they are used.

Cars, planes, telephones and ovens are all machines.

Again, a computer is just another machine – it's a device made by people to get work done.

Let's take a closer look at computers.

CHAPTER TWO
COMPUTERS

Computers were created to do a simple thing: they take in data (information), change the data in some way, and send out data. That's all.

There are certain truths regarding computers:

1. They are only machines. They are not people.

2. They were created by people and can only act if a person tells them to. Even then, they can only perform actions that a person thought of ahead of time and built into them.

Computers do not have a soul. They cannot think. Everything ever done by a computer was predetermined by humans. Even so-called "artificial intelligence" (computer systems that are able to perform actions that require human intelligence, like being able to recognize sounds and images), or computers that can "learn," only have these abilities because we designed them that way.

As machines, some of the characteristics of computers include the following:

*   They handle data. Data is information – such as words, symbols (something written that represents an amount, idea or word), pictures, etc.

*   They obey instructions (commands entered into them that perform certain tasks).

*   They automate (perform actions without human interaction) tasks that would either take too long for a person to do or be too boring. Keep in mind that these automatic actions were designed by a person.

*   They process data. "Process" means to handle something through use of an established (and usually routine) set of procedures. When a computer displays the word "processing", it is saying, "Hold on while I perform some pre-established procedures." Processing refers to "taking actions with data". Searching through words to locate typos would be an example of "processing data".

When data is being processed by a computer, you sometimes see a "progress bar" (a symbol that shows how far along something is) like this:

Or you may see this symbol when data is processed:

This circular symbol is called a "throbber" due to the fact that they originally expanded and contracted in size – i.e., the symbol "throbbed".

The purpose of computers is to take in data, process it and send it out.

When computers perform actions, it is referred to as "executing" or "running." For example, you can run a search on the internet by clicking the search button or you could execute an instruction by pressing enter on your keyboard.

It is important to understand that machines are not life forms. Even though they can perform seemingly miraculous operations, the true source of their products is humankind.

Computers aren't very useful without programs – so let's go over what programs are exactly.

CHAPTER THREE
PROGRAMS

As a noun (thing), code is the written instructions you type into a computer to make programs.

Programs are series of written instructions, entered into a computer, which cause the computer to perform a specific task or tasks. For example, Microsoft® Paint is a program that can be used to create basic illustrations on your computer.

Hardware refers to the physical components of the computer – the parts that can be touched. "Hard" means "solid" or "tangible" (able to be seen and felt), and "ware" means "something created".

The computer screen, mouse, printer and keyboard are all hardware.

The opposite of hardware is software. Software is just another word for "computer program" – sets of instructions that tell a computer what to do. Computer games, like Solitaire, are examples of software.

Another word for computer program or software is "application." These terms are all interchangeable. Though "app" (abbreviation of application) is usually used to refer to programs (applications; software) on a mobile device (like a cell phone).

The word software came about in the 1960s to emphasize the difference between it and computer hardware. Software (programs; applications) are the instructions, while hardware implements the instructions.

Installing a program means to put it into a computer so that the program can execute.

For example, you could install a program by transferring the data from the internet to your computer. In the 1990s, most software was installed using a disk that came in a box and looked something like this:

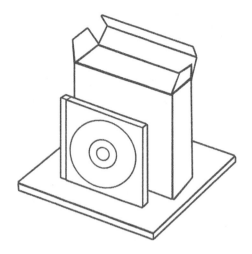

The people that create programs are called computer programmers. They're also referred to as coders, software developers, software engineers and just developers. These position names all mean the same thing.

Programs are created using programming languages, so let's discuss what those are.

CHAPTER FOUR
PROGRAMMING LANGUAGES

Languages are communication systems that allow you to transfer ideas in written and spoken words. Similar to spoken languages that people use (like Spanish and English), there are different languages that can be used to write (create) computer programs – these are called programming languages or computer languages. Programming languages are organized systems of words, phrases and symbols that let you create programs.

There are many different types of programming languages, each of which was created to fill a specific purpose. Usually a language is created in order to make the creation of certain types of computer programs easier.

As an example, Python is a popular programming language. This is how you would tell a computer to display the words "Hello, world!" on the screen, using the computer language Python:

```
print ("Hello, world!")
```

This is code. All software/applications/programs are made up of code. As a user (the person using something), you don't see the code. The three most common types of programs that developers use to write their code in are:

1) Integrated development environment (IDE): An IDE is a set of programming tools for writing software programs. IDEs are a great aid to computer program creation. An IDE often combines many available tools into one place. Essentially, an IDE is software that helps you make software.

   For example, Visual Studio, available from the technology company Microsoft, is one of the many IDEs available for software developers.

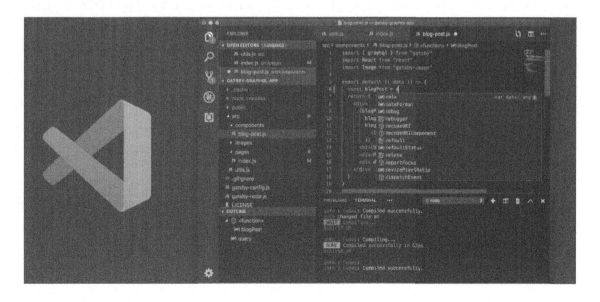

IDEs are the most common programs used by developers to write their code.

2) Text editor: A text editor is a program used to write and edit text. Text editors are very basic – meaning, the text is typically plain with no effects. This is technically different from a word processor, which is a program on a computer that allows you to create, format (design the style and layout), modify, and print documents. While text editors can be used to write code, word processors cannot. Instead, word processors have more functionality (ability to perform a wider array of actions) than text editors.

As an example of each, Microsoft Word (on the left) is a word processor, while Notepad (on the right) is a text editor.

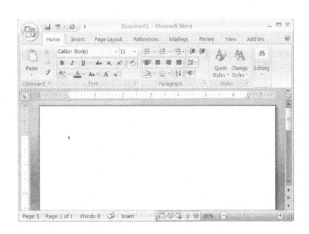 

Out of the three programs listed here used to write code within, text editors are the least popular and least recommended.

3) Code editor: As the name states, this is a program that can be used to write/edit your code. Code editors are in between IDEs and text editors in terms of features (functions built into an application) – they have fewer features than an IDE but more than a text editor. One of the most popular code editors is Notepad++ (plus plus).

As an example, the popular IDE Visual Studio has a feature called "LiveShare," which allows developers to share their code with others as it's being written. Other developers can even use LiveShare within Visual Studio to access and edit your code from a separate computer! This feature is not included in the code editor Notepad++.

On the other hand, Notepad++ has a cool feature called "auto-completion" that suggests various options for completing the code you're typing for you. For example, when you type *pr*, auto-completion may suggest *print*. This feature doesn't exist in the text editor Notepad.

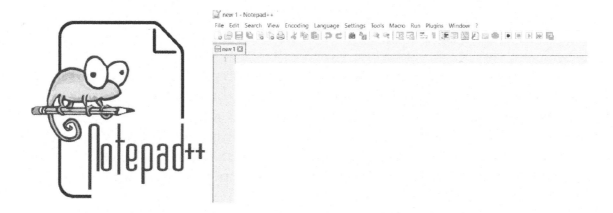

As a verb (action), coding means typing out instructions (using a particular programming language) to make a program that will make the computer perform certain actions. It also refers to the creation of websites. Developers usually code in an IDE, code editor or text editor.

Everything in computers comes down to 1s and 0s. Let's see how and why that is.

CHAPTER FIVE
MACHINE LANGUAGE

You may have heard the concept that computers operate on 1s and 0s. How this works exactly is beyond the scope of this book, but a simplified explanation is that computers are made up of billions of tiny parts that can either be on or off. 1 represents on, 0 represents off. You can instruct computers to turn these parts on and off by entering 1 and 0 into the computer. These 1s and 0s are called machine language. Commands written in machine language are called machine instructions.

As an example, "`10010100 01101011 01011011 01010110`" could instruct the computer to "`Delete the file called Vacation.doc.`" The "`10010100 01101011 01011011 01010110`" is a machine instruction.

It is difficult for people to read and write in machine language, since the instructions are just unique patterns of ones and zeroes. That is why programming languages were created. The computer converts (translates) the code written by developers into machine instructions.

It works like this:

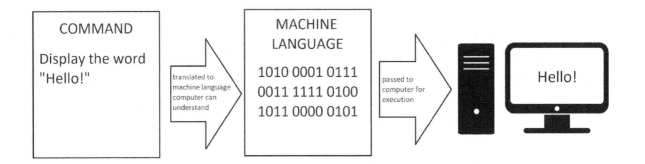

One of the main uses of computers is the internet. So, what is the internet and how does JavaScript relate to it?

CHAPTER SIX
THE INTERNET AND WORLD WIDE WEB

JavaScript is a programming language that is able to work on most computers. It was created in 1995 by an American named Brendan Eich and is particularly useful in making websites.

The World Wide Web (abbreviated "www" or the "web") is a collection of linked electronic documents, organized into groups called websites.

A website is composed of one or more individual webpages, where a "page" is an organized, separate document containing text, images, video and other elements.

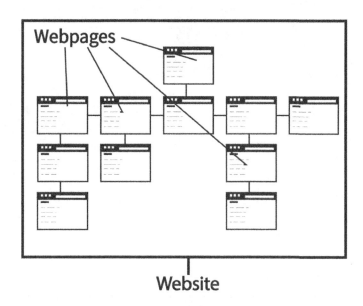

THE INTERNET

A network is a system where two or more computers are connected to each other. The computers can be connected by a cable (i.e., a wired connection) or connected wirelessly (through the air). Network is the word used to describe the link between things that are working together and is used in many different ways with computers. Information can be shared from computer to computer through the use of a network.

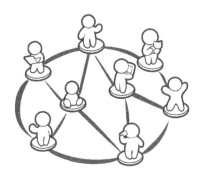

Internet is a combination of the words "interconnected" and "network."

The internet is an interconnected network of many computers around the world. It is the largest network in existence and allows computers to pass data to one another.

There are lots of different types of data that can be sent back and forth between computers connected to the internet – like electronic messages, electronic documents, healthcare records, etc.

In addition to referring to the connection between computers on this network, internet also means the set of agreements, or protocols, for how to transfer different types of data between those computers.

A "protocol" is an official procedure. In technology, it is a formal description of how a certain type of information will be formatted and handled. Basically, it's an agreement that the various people who work with that type of information all adhere to. Protocols are usually described in written documents, and are very precise. They are usually created by experts in the applicable industry.

An example of a type of data where a protocol would be valuable is healthcare information. If various organizations in the healthcare industry were to transfer healthcare data back and forth between computers as they perform their work, it would be important that they all agree about things like the exact format of the information, how to keep private data safe, etc. All of that would be laid down in a written protocol. Several such protocols do exist in the healthcare industry.

One or more protocols have been created for each type of data that can be transferred around on the internet. Violation of these protocols results in an error and the data will not be sent properly or at all.

So to recap: the internet is an interconnected network of many computers around the world, and a set of agreements, or protocols, for transferring different types of data between those computers.

INTERNET VS. WORLD WIDE WEB

So, what's the difference between the internet and the World Wide Web? The internet is the hardware and protocols that data is sent across, whereas the web is one type of information that is accessed over the internet. The web is a collection of linked electronic documents called webpages.

The internet is the infrastructure (physical framework) while the web is the code that is transmitted and displayed. In a way, the internet is the hardware and the web is the software.

Dynamic refers to actions that take place the moment they are needed, rather than in advance. For example, a restaurant that prepares your food to your specifications when you order, would be dynamic. The opposite of dynamic is "static," which literally means, "unchanging." A static restaurant would have food pre-cooked and waiting before you order – the food doesn't change in regards to time or circumstance.

In coding, dynamic means the computer processes your request right when you ask it to – it does not have a predetermined result ahead of time.

Dynamic relates to the internet by describing websites that have content that changes while the website is being viewed.

These are mouse pointers:

As an example, a dynamic picture on a website could be one that has an effect (movement or change) when you hover over it with a mouse (hovering with a mouse refers to holding the pointer over a particular object).

PROGRAMMING

Programming (the action of creating computer programs by writing code; coding) is a spectacular thing because it is one of the few skills that apply to virtually all industries.

Yes, companies that create software definitely utilize coding the most – but if you think about it, most industries utilize technology (software, websites, databases, etc.). And so, coders are valuable assets for companies in any industry – construction, food service, retail, transportation, etc.

There are many, many programming languages, and technology is ever-changing. The whole thing can be quite overwhelming, but there are basics that apply across the vast landscape of technology.

Now, let's discuss what Javascript is.

CHAPTER SEVEN
JAVASCRIPT

JavaScript is a great language to use for an introduction to coding and tops some lists as the most popular programming language in the world.

One of the main uses of JavaScript is to make websites more dynamic (describing websites that have content that changes based on user action or other factors, rather than being static in appearance and content).

JavaScript can also make programs and websites more interactive, such as making it so videos start to play as soon as a user moves their mouse over them.

There are many other uses for JavaScript; it is used in the creation of many different types of computer programs.

So, what is the definition of JavaScript? Well, according to one of the top online technical dictionaries, "JavaScript is a high-level, untyped, and interpreted runtime language."

Right off the bat, that explanation shows why we made this book. That definition makes no sense. So, in order to comprehend what is being said there, let's dissect the definition.

HIGH-LEVEL

Machine language that we discussed earlier (1s and 0s) is considered a "low-level language." This is due to the fact that it is close to the hardware of the computer in that it directly represents the on or off states used in both the computer instructions and the data the computer operates on.

The opposite of a low-level language is a high-level language – code written in a way that is easier to read and write because it somewhat resembles English, such as:

```
Print "Hello";
```

Here, you are telling the computer to write out the word Hello on the screen.

All the most popular programming languages are high-level languages – meaning, they use English letters, symbols and words.

<u>UNTYPED</u>

Data is information.

There are different types of data that computer programs will need. These are data types like:
- Numbers,
- Letters,
- Dates,
- True or false, etc.

When information is used in a computer, that information will have a "data type." This is two things:

1. "What kind of information is this?," and

2. "What kind of things can we do with this data?"

Different data types can be used in different ways.

When data is stored in a computer, its type is also stored – that way the computer knows how to work with that particular piece of data.

The various different things you can do with data are called operations. Not all data types have the same operations available to them.

For example, a "decimal number" (0-9) is a data type. Typical operations you could do with decimal number data are addition, subtraction, multiplication, etc.

"Text" is another data type. Typical operations you could do with text data are "convert the text to uppercase" or "concatenate" (which means to link together, as in a chain – for example, you could take the individual text elements "Pizza" and "s!" and then concatenate them into "Pizzas!" etc.).

When creating computer programs, every piece of data being kept track of by the computer is a certain data type. For example, the datum 342.98 is data of type "decimal." The datum "casserole" is data of type "text."

A B C   1 2 3   TRUE/FALSE   3.14
(examples of data types)

It would not make sense to apply mathematical operations to text data – for example, trying to tell a computer to add the number 34 to the text "banana" wouldn't work because the two types of data are different, and mathematics doesn't apply to text data.

In math, a variable is a symbol used to represent an unknown quantity or a quantity that may change. In computers, a variable is a construct (an idea formed from simpler elements) used to store data that may change as the computer performs its tasks.

Computers often have to keep track of various pieces of information. These can be things like the name of a computer user, the color of the background shown on the computer screen, an order number for something a user ordered from a company, etc.

The computer usually gives each of these pieces of data a NAME and a VALUE. The NAME is used to identify the exact piece of data, and the VALUE is used to show the actual data we need to keep track of. Usually the "=" symbol is used to set the VALUE of a variable.

In math, a value is a numeric amount, quantity, or number – like this: `X = 10`

Here we are saying that the *name* "X" holds the *value* of "10."

In coding, a value is any sort of characteristic or attribute. We could assign the value like this:

```
X = Happy
```

This means we are saying the *name* X is assigned the *value* Happy. We could later tell the computer:

```
Print X
```

The computer would then display:

```
Happy
```

The things the computer is keeping track of are usually called "variables." This is because the VALUE part of it can change when you tell the computer to change it. The fact that the value can vary is why it's called a "variable" – its value is not permanent or constant.

Many computer programming languages require the programmer to specify the data type for a variable. These are called "typed languages." When you first create a variable in a typed language, you need to specify what data type the variable is.

In a typed language, there are certain restrictions about how the various data types can interact.

For example, you might need to make the computer instructions that calculate a person's pay for the day. Here, you could have three variables that had a data type of "integer," meaning they represented numeric data (i.e., numbers). An integer is a whole number – like 6, 99 and 182. 3.11 and 9½ are not integers.

You would create the variables that calculate a person's pay for the day like this:

```
integer PayRate = 35.00
integer HoursWorked = 10
integer Paycheck
```

The third variable, the integer called Paycheck, doesn't have a value yet. This variable is meant to store the result of a math operation that determines how much a worker has earned.

The computer instruction for this might look something like this:

```
Paycheck = HoursWorked x PayRate
```

Because all three variables are the same data type, the computer can process this math operation and store the result in the value for variable "Paycheck": 350.00.

Untyped languages are programming languages that do not make you define what data type a variable is. Instead, the computer uses a predefined system to work out what data type a variable is based on its value and how it is being used.

For example, let's use the earlier math example but examine how it could work when performed using an untyped language.

Here, we might have three variables similar to the ones from our earlier example:

```
PayRate = 35.00
HoursWorked = "ten"
Paycheck
```

Notice that there are no data types specified. Notice, as well, that the variable "HoursWork" has a value of "ten" – a word, not a number.

If we wanted to perform the same math as the above example, we might create the same computer instruction:

```
Paycheck = HoursWorked x PayRate
```

This makes math operations a bit different. The computer takes the value of the variable PayRate, 35.00, and tries to multiply it by the value of the variable HoursWorked, "ten."

This doesn't work at first glance, of course – you can't multiply a number by a word! But untyped languages have a predetermined system that is used to infer (determine something based on the available information) the type of data that the variable represents. Here, the language can work out that the word "ten" can mean the number 10 – and since the instruction is a math operation, the language interprets the variable in that manner and multiplies 35.00 by 10 to get 350.

JavaScript is an *untyped* programming language in that you aren't required to define the type of a variable.

INTERPRETED

A compiler is a special program that converts the code that looks nearly English into a form that the computer can understand and operate off of. The product of a compiler would be a set of instructions for the computer to execute that would end up doing what the "nearly English" code described.

For example, if you typed this instruction within a computer program:

```
if age > 18, print "You are an adult"
```
(the symbol > means greater than)

A compiler would take what you wrote, change it into a form the computer can understand and then relay that command to the computer so it can be executed (i.e., machine language –a long series of 1s and 0s).

One important aspect of how a compiler works is that it takes all of the instructions in the program and turns them into language the computer can understand before actually executing the program. There are other methods of converting high-level languages to computer-readable form that don't involve converting the entire program before it is executed. Instead, each instruction is converted and executed one at a time. This is not how a compiler works, however.

A compiled programming language uses a compiler.

An interpreter is a special program that converts high-level language instructions into low-level language instructions. An interpreter converts code in this way one instruction at a time rather than converting the entire set of instructions in the program prior to execution.

In other words, the interpreter will read an instruction, convert it to the language the computer can understand, have the computer execute that instruction, then go on to the next instruction in the program and repeat.

Interpreted programming languages (like JavaScript) use interpreters.

RUNTIME

Run means to start or execute something.

To understand runtime, you must first understand what a "compiler" is.

Runtime means the time when something in the computer is running. Runtime is exactly what it means – the time when the program is run. You can say something happens at runtime or it happens at compile time.

The term "runtime" also describes software or instructions that are executed while your program is running, especially those instructions that you did not write explicitly, but are necessary for the proper execution of your code. This is actually a "runtime library," but is often shortened to simply "runtime."

A library is a collection of pre-made resources used to create computer programs. Most often, these take the form of a package of code with built-in functions that all relate to a specific area. They are usually used by adding them to a software project already being created, and then having code in that project make use of the various functions in the library.

In computer programming, a runtime library is a special program library used by a compiler to implement functions built into a programming language during the runtime (execution) of a computer program.

For example, you can design videos to load during runtime or prior to runtime. If they were loaded during runtime, then you would have to wait for it to load after you clicked on it. If they were loaded prior to runtime, the videos would have loaded when you originally opened the program (set of instructions entered into a computer that performs exact functions) and so the program would take longer to load up when started.

PUTTING IT TOGETHER

So, to circle back to the definition given at the beginning of this chapter. "JavaScript" is "a high-level, un-typed, and interpreted runtime language." With the data covered in this chapter, you can now understand this definition!

Now let's go over how to solve problems in coding.

CHAPTER EIGHT
ALGORITHMS

"Algorithm" is a mathematical term that means a plan for solving a problem. It consists of a sequence of steps to solve a problem or perform an action.

Computers use algorithms. An algorithm is a set of instructions that is used to get something done.

For example, an algorithm for selecting the right kind of shirt might have the following steps:

```
1. Pick a shirt from your closet.
2. Put the shirt on.
3. Look at yourself in the mirror.
4. If you like it, go to step 6.
5. If you don't like it, return to step 2.
6. End of procedure.
```

Computers perform functions by processing algorithms. You can create algorithms for computers to follow that other people can then use in the future without having to write out the entire list of steps again.

For example, when your computer prints something, there are many massive algorithms it uses to figure out which type of printer you have, how to convey the document to the printer in a way it understands, etc. This is all taken care of in the background so that all you need to do is click "print." These algorithms were written by others to get the computer and printer to comply with your wishes.

Another example of an algorithm could be a simple recipe, such as:

```
1. Crack three eggs and pour their contents into a small bowl – attempt to
   avoid getting eggshell in the bowl.
2. In the case of eggshells, remove any eggshells from the bowl.
3. Beat the 3 eggs for 90 seconds in the small bowl.
4. Put a small frying pan on the stove and turn the burner on to medium
   temperature.
5. Pour the eggs in a small frying pan and cook until very little to no
   wetness remains.
6. Serve the eggs.
```

While there are similarities, computers and people do not use algorithms in the same way; there are similar features between real life algorithms and computer algorithms, but they are not identical.

Computers can't think – they operate in a completely literal manner – so when writing an algorithm for computers, there are certain characteristics that must be present for a computer to use the algorithm. People can deduce things and extrapolate data; people have imagination. Computers can only do what they've been told to do.

To fully understand what an algorithm is for computers, some terms and concepts must be cleared up first. Algorithms are composed of five separate characteristics:

1. Result: the outcome of something.

2. Finite amount of time: an exact amount of time for something.

3. Well-ordered: listed out in the correct sequence and in an organized fashion.

4. Effectively computable operations: this refers to a command that can be processed successfully by a computer.

5. Unambiguous: ambiguous means unclear, confusing, not specified. Un = not. Something that is unambiguous means that it is clear and defined.

Let's go over each of these points in detail.

RESULT

Result: the outcome of something.

In computers, a response could be considered the result. You type in: "Tell me the date," the computer replies: "Jan 3, 2022." "Jan 3, 2022" would be the result. Answers to problems are also considered a result.

Algorithms always have a result. If there is no result, it cannot be considered a valid algorithm.

For example: `print the number "2,"` Result: 2.

FINITE AMOUNT OF TIME

Finite amount of time: an exact amount of time for something.

Algorithms should have a finite number of operations included in them and should execute these operations in a set period of time. Algorithms should never be open-ended or infinite. Since each step of an algorithm takes a certain amount of time to be completed, this would mean the algorithm would be completed in a finite amount of time.

An incorrect algorithm would be:

```
1. Start at an amount of 1
2. Add 1 to that amount
3. Go to step 2
```

The computer would then count 0, 1, 2, 3, 4, etc. forever.

For example, "Count from 1 to 10, then print a message that says, 'I counted from 1-10'," would be a correct algorithm because it contains a finite amount of steps.

WELL-ORDERED

Well-ordered: listed out in the correct sequence and in an organized fashion. Computers can only execute an algorithm if it is well-ordered. If it is put out of sequence or no sequence is specified, etc., the computer cannot process the algorithm.

For example, this is an incorrect algorithm because it is not well-ordered:

```
1. Close the program
2. Print "I am here" inside the program
3. Save your work in the program
4. Open the program
```

EFFECTIVELY COMPUTABLE OPERATIONS

Effectively computable operations: this refers to a command that can be processed successfully by a computer. The phrase literally means an operation that can be computed in an effective way.

In algorithms, all of the parts of the algorithm must be possible to execute.

For example, a non-effectively computable operation could be: "multiply green by red" – the computer can't process this because it isn't computable. Operations contained in algorithms must be effectively computable operations, such as, "2 + 2 =".

UNAMBIGUOUS

Unambiguous: ambiguous means unclear, confusing, not specified. Un = not. Something that is unambiguous means that it is clear and defined.

In computers, your algorithms must be unambiguous. You have to be extremely specific about each step/part of an algorithm or the algorithms cannot be processed by the computer.

For example, if you have an algorithm that is processing a list of ten numbers, then a step of your algorithm is "delete the number," the computer cannot process this because you did not clearly specify which number you wanted deleted.

Now that you know the above terms and concepts, we have the full definition of "algorithm" as defined in the book *Invitation to Computer Science* by American authors Schneider and Gersting:

Algorithm: *A well-ordered collection of unambiguous and effectively computable operations that when executed produces a result and halts in a finite amount of time.*

Sorting means to put items in an exact sequence or order. Computers sort items often. On your computer, you can sort files by when they were created, their name, etc. Algorithms can be used by computers to organize data.

For example, you can sort files on your computer by name or date. This is done by a sorting algorithm within the computer. There are several different ways algorithms can be written to sort data.

In learning how to program, you would write algorithms to perform various functions. Computers have many algorithms programmed into them already that cause them to do a lot of the actions you use computers for. For example, your computer performs many complicated steps to save a document on your computer. These steps are taken care of because programmers entered in algorithms to handle it for you.

Whether or not you ever learn to write code, it is helpful to know the elements that make up a program, which we will now cover.

CHAPTER NINE
FIVE ELEMENTS OF A PROGRAM

There are five key elements to any computer program:

1. Entrance
2. Control/Branching (decision points within a program)
3. Variables
4. Subprograms (programs within programs)
5. Exit

Let's go over each of these in detail.

ENTRANCE

A computer is a simple machine when you get down to it. It can only do one thing at a time, and it performs a computer program's instructions in the exact order in which the computer programmer puts them. It can only execute (perform or run) an instruction if it is directed to.

This means that any computer program has to have a clearly marked "first instruction." This is the first task that the computer will perform when the computer program is started. From that point forward, each instruction in the program will direct the computer what instruction to perform next after it performs the current instruction.

There are different ways to specify the entrance point, depending on which computer programming language is being used, but every computer program has a defined entrance point.

CONTROL/BRANCHING

Computers are often used to automate actions that would otherwise be performed by people. One of the most common things a person will be asked to do in performing a job is to assess the condition of a thing and, based on the condition of that thing, choose between two or more possible courses of action. In other words, they will have to make a decision. An example would be the activity of "a teacher grading a stack of papers":

1. Take the next student paper from the top of the stack.
2. Grade the paper.
3. Write the grade on the paper.
4. If the grade is 70% or higher, put the paper in a "Passed" stack.
5. If the grade is below 70%, put the paper in a "Failed" stack.

You can see that there are two possible "paths" here. A path is "a possible course of action arising from a decision." Think of it as what happens when you come to a fork in the road. You have to decide on a course of action – which road do you take? This is also called a branch.

All but the simplest of computer programs will need to do the same thing. That is, they will have to check the condition of a piece of data, and based on the condition of that data, they will have to execute different sets of computer instructions.

In order to do this, the program will make use of special computer instructions called "control" instructions. These are just instructions that tell the computer what to look at in making a decision, and then tell the computer what to do for each possible decision. The most fundamental control statement for a computer is "if." It is used like this:

```
IF [condition to check] THEN [branch of computer instructions to execute]
```

Here, the "IF" is the control statement; the "THEN" is the branching instruction that points to the path of the program to execute if the control statement is true.

VARIABLES

We covered variables earlier and they are a key part of programs.

As a recap, a variable is a piece of data that a computer program uses to keep track of values that can change as the program is executed. This might be something like "the grade of the student paper that was just graded" or "the color of paint to use for the next car on the assembly line."

Variables are a vital part of any computer program because they make it so a computer program can be used for more than a single, predetermined set of values. You can imagine that if "the color of paint to use for the next car on the assembly line" was only ever able to be "blue," the computer program using that data wouldn't be very useful. It would make a lot more sense to make it so the computer program could change that value for each car that was going to be painted.

When you are writing variables in a computer program, they usually are written in a manner like this:

```
[name of the variable] = [value of the variable]
```

For example, you might have something like this:

```
color = "red"
```

Here, the variable is named "color," and the value of that variable has been set to "red." In other words, the variable named "color" is now "equal" to the word "red."

Let's look at the example of "a teacher grading a stack of papers." Here, we could have a variable called "Paper Grade" that changed each time the teacher graded a paper. You could also have variables for the total number of questions on the paper ("Total Questions") for both the number of questions the student answered correctly ("Correct Questions") and for the grade of the paper.

The written description from above:

```
1. Take the next student paper from the top of the "To Be Graded" stack.
2. Grade the paper.
3. Write the grade on the paper.
4. If the grade is 70% or higher, put the paper in a "Passed" stack.
5. If the grade is below 70%, put the paper in a "Failed" stack.
```

In computer language, the procedure might look something like this:

```
1. Retrieve Next Paper
2. Set Total Questions = [total questions in current Paper]
3. Grade Paper
4. Set Correct Questions = [number of questions answered correctly]
5. Set Paper Grade = [Correct Questions/Total Questions]
6. If (Paper Grade >= 70%) then Paper Status = "passed"
7. (Paper Grade < 70%) then Paper Status = "failed"
```

This is a simple computer program.

As a note, the >= (greater-than sign followed by equal sign) is a symbol used to show that a comparison should be made. Specifically, this "greater-than or equal" symbol is an instruction to check whether the data on the left side of the symbol is more than or equal in amount or quantity to the data on the right side. The answer to this comparison is an answer of "true" or "false."

For example: 5 >= 4

This means "check whether 5 is greater than or equal to 4." Since five is in fact greater than four, the answer is "true."

Another example: 3 >= 6

This means "check whether 3 is greater than or equal to 6." The answer is "false."

As a final example: 6 >= 6

This means "check whether 6 is greater than or equal to 6." The answer is "true."

The reverse (opposite) symbol of >= is <=, which checks for whether the data on the left side of the symbol is *lesser/fewer* or *equal* to the data on the right side.

Each time the computer runs the seven-step procedure listed earlier, it could have different values for each of the variables in the program, depending on how many questions the paper being graded has and how many of those questions the student answered correctly.

For example, let's say the paper has 100 questions, and the student answers 82 of them correctly. After the program is run, the result would be the following:

```
Total Questions:      100
Correct Questions:    82
Paper Grade:          82%
Paper Status:         Passed
```

As another example, let's say the paper has 50 questions, and the student answers 30 of them correctly. After the program is run, the result would be the following:

```
Total Questions:      50
Correct Questions:    30
Paper Grade:          60%
Paper Status:         Failed
```

To clarify the need for variables: Let's say that at the time this computer program was being created, all papers at the school had 100 questions, and the teachers told the programmer to make it so that the number of questions was always assumed to be 100. In that case, the programmer wouldn't use a variable called "Total Questions."

Instead, they could make the program look like this:

```
1. Retrieve Next Paper
2. Grade Paper
3. Set Correct Questions = [number of questions answered correctly]
```

```
4. Set Paper Grade = [Correct Questions/100]
5. If (Paper Grade >= 70%) then Paper Status = "passed"
6. If (Paper Grade < 70%) then Paper Status = "failed"
```

Notice that on line 4 of the program, the programmer set the number of questions to 100.

Now, let's say that the school introduces the concept of "quizzes," which are smaller papers with only 20 questions. If the paper being handled by the computer program is a quiz, the grade will no longer be accurate – even if a student got all 20 questions correct, they would only get a grade of 20% (20/100).

A good programmer will analyze the need that the program is meant to resolve, then build the program so that it can handle changing aspects of that need over time.

Another valuable control statement is a loop. This is where part of the program is executed over and over until a certain condition is met.

In real-world terms, an example might be "grade papers one at a time until all the papers have been graded" or "make five copies of this document."

In a computer program, a loop would look something like this:

```
• [start loop]
    o Perform action
    o If [end condition has been met] then [exit the loop]
    o If [end condition has not been met] then [repeat the loop]
• [end loop]
```

The program we looked at that grades papers could be set up as a loop. The instructions would be laid out like this:

```
• [start loop]
    o Take the next student paper from the top of the "To Be Graded" stack.
    o Grade the paper.
    o Write the grade on the paper.
    o If the grade is 70% or higher, put the paper in a "Passed" stack.
    o If the grade is below 70%, put the paper in a "Failed" stack.
    o If there are no more papers in the "To Be Graded" stack, exit the loop.
    o If there are more papers in the "To Be Graded" stack, repeat the loop.
• [end loop]
```

Often loops make use of a special variable called a "counter." The counter keeps track of how many times the loop has been executed. This can be used to make sure the loop is only executed when needed.

Let's add a counter to the grading program we're looking at as well as two new variables: "Total Papers" will be used to hold the value "how many papers need to be graded," while "Counter" will be used to hold the value "how many times the loop has been executed."

```
1. Set Total Papers = [total papers to be graded]

2. Set Counter = 0

3. If (Counter < Total Papers):
   a. Retrieve next Paper
   b. Set Total Questions = [total questions in current Paper]
   c. Grade paper
   d. Set Correct Questions = [number of questions answered correctly]
   e. Set Paper Grade = [Correct Questions/Total Questions]
   f. If (Paper Grade >= 70%) then Paper Status = "passed"
   g. If (Paper Grade < 70%) then Paper Status = "failed"
   h. Counter = Counter + 1
   i. Go to step 3

4. [Continue on with the rest of the program]
```

Here, the loop is found in step 3.

Let's break down what each step is doing here:

Step 1:   Count how many papers are in the "to be graded" stack and set the value of the "Total Papers" variable to that number.

Step 2:   Create a variable called "Counter" and set it to the value zero. This variable will be used to keep track of how many papers are graded.

Step 3:   Use the control statement "if" to see if we should execute a loop.

Step 3a–3g: Grade the paper; this has been covered above.

Step 3h: Since we have now graded a paper, add one to our Counter variable.

Step 3i: Go to the top of the loop, where we check to see if we need to execute the loop all over again.

Let's see what would happen if we used this program to grade two papers. Let's say that the papers look like this:

```
Paper 1:
Total questions on the paper: 100
Total questions that were answered correctly: 95

Paper 2:
Total questions on the paper: 20
Total questions that were answered correctly: 10
```

If we analyze what happens when the program is executed by the computer, it would look like this:

```
Total Papers = 2
Counter = 0
0 is less than 2, so loop will be executed
Paper 1 Retrieved
Total Questions = 100
Paper 1 Graded
Correct Questions = 95
Paper Grade = 95%
Paper Status = "passed"
Counter = 1
1 is less than 2, so loop will be executed
Paper 2 Retrieved
Total Questions = 20
Paper 1 Graded
Correct Questions = 10
Paper Grade = 50%
Paper Status = "failed"
Counter = 2
2 is not less than 2, so loop will not be executed
[Continue on with the rest of the program]
```

SUBPROGRAMS

As covered earlier, computer programs are generally executed in order, from the start point to the end point. This is called the "path of execution."

The main series of instructions in a program is called the "main program."

It is sometimes valuable to create another program that can be used by the main program as needed. This is called a subprogram. It is no different from any other program – it is made up of the same elements (entrance point, variables, control and branching statements, and exit point).

However, a subprogram isn't used all by itself. Instead, the main program can execute the subprogram as needed. Here, the main program stops executing, and the subprogram starts executing. When the subprogram is done executing, the main program continues on where it left off.

This is referred to as "calling" the subprogram – that is, the main program calls the subprogram, the subprogram starts and stops, and the main program continues on where it left off before calling the subprogram.

This is useful in creating programs because the computer programmer doesn't have to enter the instructions of the subprogram over and over. You only type them in once, and then when you need that subprogram to be called by the main program, you only have to type in one instruction – the instruction to call the subprogram. This lets you reuse the instructions you entered in for the subprogram rather than rewriting them.

EXIT

Every program must have an instruction that tells the computer that the program is no longer running. Much like the Entrance, the exact instruction for this varies based on the computer language used, but all computer languages will have this type of instruction.

SUMMARY

The fundamental elements of any computer program are:

1) An entrance point – the first instruction to be performed.

2) Control and branching statements – to control what is done and in what order.

3) Variables – changeable items, held in computer memory, for the program to use as it operates.

4) Subprograms – repeatable sets of computer instructions that act like mini-programs. The main program can use them as needed.

5) An exit point – the last instruction to be completed, so the computer knows the program isn't operating anymore.

Next we will discuss types of programming languages.

CHAPTER TEN
TYPES OF PROGRAMMING LANGUAGES

As we covered earlier in this book, programming languages are organized systems of words, phrases and symbols that let you create programs. They're also called computer languages.

There are many different types of programming languages, each of which was created to fill a specific purpose. Usually a language is created in order to make the creation of certain types of computer programs easier.

Some programming languages are specifically made to create websites, while others are solely for making apps. In this chapter we are going to cover some of the most well-known programming languages. But first, there are a couple of terms to cover in relation to programming languages. The first is "markup."

"Markup" is the action of adding instructions to a document to control the format, style and appearance of the content of the document. This applies to physical printed documents, as well as to electronic documents that are displayed on a computer screen (like websites).

A markup language is a type of programming language that is used to specify markup instructions for the data in a document. In other words, markup instructions aren't the data in the document; they are data *about* the data in the document.

There are several types of markup languages. Usually they were created for commonly-used types of documents which have exact specifications as far as their structure. Often, these specifications as to document structure are well-known around the world and available to anyone who wants to create documents of that type.

A script is a list of commands the computer performs automatically without your involvement. A scripting language is a computer language used to make scripts.

Often the tasks these scripts accomplish are valuable to automate – so that the task steps don't have to be entered into the computer again every time you want to do those tasks.

The origin of the term is similar to its meaning in "a movie script tells actors what to do"; a scripting language controls the operation of a computer, giving it a sequence of work to do all in one batch. For instance, one could put a series of editing commands in a file, and tell an editor to run that "script" as if those commands had been typed by hand.

As an example, if you wanted to make it so that every night at midnight, a list of all the people who signed up for your school classes was printed, you could use a scripting language to make a script that did this work.

JavaScript can be used to create scripts and is a scripting language.

There are a couple of languages used with JavaScript – let's go over what those are.

CHAPTER ELEVEN
HTML AND CSS

JavaScript is one of three of the main technologies utilized to create content on the world wide web (HTML and CSS being the other two).

HTML

HyperText Markup Language is the markup language that makes up the webpages on the World Wide Web. It was created by Tim Berners-Lee (the inventor of the web) in 1990. Hypertext is a system for linking electronic documents. In an HTML document, hypertext is words, images, videos, etc. that link to other HTML documents. When a user selects the hypertext, the linked document is retrieved and displayed on the computer screen. HTML is used to create webpages.

A computer programmer uses HTML to put markup instructions before and after the content of a webpage. These markup instructions give the computer data *about* the elements of the document – text, images, video, etc. When the document is actually displayed on a screen, those markup instructions aren't displayed on the screen – instead, they are used to control *how* the actual content of the document is displayed. As of 2020, HTML5 was the newest version of HTML. Here is an example of an HTML website from the 1990s:

CSS

Cascading Style Sheets was created in 1996 and is used to control the appearance of the text, images, videos, etc., on those webpages. A style sheet is a tool used to design the format of a webpage. Style sheets are a central place to store data about how that page will appear and how it will behave. They can be used to store information about how to display colors, pictures, text, etc. A style sheet is a sheet that handles the style of the website.

Styling is not concerned with the content of a page; rather, it is concerned with the appearance of that content. Cascading is a term that comes from the idea of water dropping down a series of drops in height. Think of it as a series of small waterfalls. Information can cascade, meaning it can be stored at one central, most important location and then can be applied to other locations that are considered to be lower than, or derived from, that more important central location. For web pages, that means you can set style information in one high-level web page, and that same styling information can "cascade" to other, lower pages without any further work. CSS is usually used in conjunction with HTML to upgrade the appearance and functionality of HTML websites. As of 2020, CSS3 was the newest version of CSS.

HTML, CSS, and JavaScript are commonly used in conjunction with each other to create dynamic websites. In this book, we will teach you some basic JavaScript, and to do so, we will use HTML occasionally.

HISTORY OF JAVASCRIPT

A browser is a program that you use to look at webpages on the internet. In order to look at information on the web, you use a web browser. The most common activity people use browsers for is finding and viewing websites – more particularly, the webpages on websites. Google Chrome is the most used browser.

The difference between a browser and search engine is: a browser is used to visit and display websites, while a search engine is used to search for websites. For example, you could use the browser Internet Explorer to search on Yahoo's search engine.

In the mid-1990s, there was no Google, and so, prior to Google Chrome, the most popular browser was Netscape Navigator. JavaScript was created by Brendan Eich in 1995 who worked for Netscape at the time.

Java was a popular programming language at the time, and still is – and was one of the inspirations for JavaScript. But to be clear, Java and JavaScript are two completely different programming languages. Since Java was being heavily marketed at the time, Netscape named the new scripting language JavaScript.

VALUE TYPES

Nearly every computer programming language comes with certain data types built in and allows for the creation of other types of data by computer programmers. In JavaScript, these various types of data are called "values."

There are six value types in JavaScript:

1. Strings (data that represent text or a series of text)

2. Numbers

3. Booleans (true or false statements)

4. Objects (things with state and behavior)

5. Functions (a block of organized, reusable computer code used to perform a single, related action – also called a subprogram or subroutine)

6. Values that are undefined

You will learn how to utilize these value types in this book.

We are really close to writing code! Let's go over some tips to help you through this book.

CHAPTER TWELVE
CODING TIPS

If you run into any trouble while going through this book, here are some tips:

1) Ensure you understand all the words and terms being used – clear up any you don't understand.

2) Ensure your code is written exactly as laid out here. A small error in the code, such as a missing comma, can ruin the whole program. Code must be exact for programs to run properly, so always meticulously check your code for errors.

3) Research online for solutions.

4) You can also contact The Tech Academy and ask for assistance – learncodinganywhere.com

CHALLENGES

At the end of some chapters, we will have an "END OF CHAPTER CHALLENGE." These are opportunities for you to put together all that you've studied in that chapter. At times you will also be instructed to figure out solutions to problems on your own. Working software developers are often assigned projects and tasks they've never done before. A key element of the job is researching solutions online. You'll find that some of the challenges in this book will instruct you to do something that we haven't taught you yet. This was done intentionally. We want you to gain experience in locating new data online and figuring out things on your own. In the words of more than one software developer: "I get paid to Google things!"

Some of these challenges will be a repeat of tasks you've already performed. The reasons for this are to:

a. Provide you with an opportunity to create your own approach, and

b. Allow you to better understand and remember code through repetition.

Well done on making it through these beginning chapters! Now we are going to start coding!

CHAPTER THIRTEEN
FIRST CODING ASSIGNMENTS

Here is your first task:

```
Download and install Notepad++ here: https://notepad-plus-plus.org/
```

You will be writing your code in the code editor Notepad++.

Note: If you are using a Macintosh computer, you can utilize a different code editor – such as Brackets.

HTML

Before writing code in JavaScript, you must be familiar with some basic HTML.

The markup instructions in HTML give the computer data *about* the elements of the document – text, images, video, etc. When the document is actually displayed on a screen, those markup instructions aren't displayed on the screen; instead, they are used to control *how* the actual content of the document is displayed.

TAGS

HTML utilizes tags. Tags are how the markup of the content in the HTML document is specified.

The basic principle is this: If you put special instructions before and after the content you want to affect, another program can tell the difference between the instruction and the content, and can then present the data in a specified manner.

Because of this, tags usually have two parts: an opening tag and a closing tag. The content would go in between these two tags. Opening tags indicate the beginning of a markup instruction and closing tags indicate the end.

Opening tags are written as a "less than" angle bracket (<), followed by the markup instruction, and then a "greater than" angle bracket (>).

An example would be the tag used to make text **bold**. The actual instruction is the word "strong", meaning "make the text controlled by this tag stand out", usually by making it bold. The opening "strong" tag would look like this:

```
<strong>
```

You need to specify a closing tag, so the computer program knows when to stop making text bold. That's where the closing tag comes in. Closing tags use the same instruction as the opening tag, but with a forward slash (/) placed before the instruction (</>).

The closing "strong" tag would look like this:

```
</strong>
```

Putting all the parts together, if you wanted to emphasize a quote, you would write the HTML like this:

```
<strong>"Education is the most powerful weapon which you can use to change the world."</strong> -Nelson Mandela
```

The output of this code would look like this:

"Education is the most powerful weapon which you can use to change the world."
-Nelson Mandela

ELEMENTS

The tags and the code written between them are called "elements." Here is a diagram that shows this:

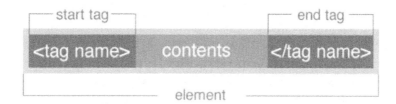

In this chapter, we will be writing HTML and JavaScript together. That way, you can see how they work together. Additionally, when you state that you know JavaScript, it will be assumed that you also know HTML.

NOTEPAD++

Now that Notepad++ is installed on your machine, do the following:

```
Launch Notepad++
```

```
Click on File on the menu bar and open a New File (or just press Ctrl + N).
```

"HELLO, WORLD" IN HTML

The first code that developers usually write when learning a new language is to make the computer display the text: "Hello world." We are going to first use HTML to do this.

At the top of each HTML document containing HTML code is the <!DOCTYPE HTML> declaration. This isn't a tag. It is an instruction to the browser telling it you're using HTML – there are other doctypes (types of documents) you could use, so clarifying that helps the browser out.

The <html> tag is inserted at the beginning of your code, and all your HTML code is contained in between it and the end tag (</html>).

The <body> tag denotes where you can add the contents of an HTML document – such as text, images, videos, lists, links, etc.

Inside Notepad++, write the following code:

```
<!DOCTYPE HTML>
<html>
    <body>
    Hello, world!
    </body>
</html>
```

As a reminder, the "/" symbol in the code you wrote denotes an "end tag."

Save this document by clicking on "File," then "Save As." Name the document Test.html and ensure you click on "Hypertext Markup Language" under "Save as type." It is important to include .html at the end of all of your HTML code files because if you don't, your computer will not recognize that it is an HTML file and your code won't run. Also, make sure you save this file to your desktop.

There are various ways to run your code. One of the easiest is to drag and drop the HTML file into an open Google Chrome window, like this:

Note that the file icon (symbol above the file name) in the picture above is the Google Chrome logo, that's because on the computer the photo is taken from, Google Chrome is set as the default program used to open HTML files. The icon of your file may or may not be different.

You can also right click on your file on the desktop and select "Open with" and navigate to Google Chrome.

From now on in this book, when we say "run" or "execute" your code, we mean to "Open it in Google Chrome" so you can see your webpage displayed in the browser.

For Mac users using Brackets, you can also click "File" and then "Live Preview" to run your code.

Now do the following:

```
Run your code.
```

Congratulations! You made the browser display the text "Hello, world!"

ONLINE IMAGES

A Uniform Resource Locator (URL) is a unique name for a web resource. A "web resource" is an identifiable thing that can be found on the internet. The most common example of a URL is a website – https://www.google.com is a URL.

An attribute is a word or phrase used inside an element's opening tag that controls the element's behavior.

For example, the image <img> tag allows you to place images inside your webpage. The src (source) attribute is used with this <img> tag and it gives the location of the image (where the image is being displayed from).

You can see this in the following:

```
<img src="https://www.petmd.com/sites/default/files/petmd-cat-happy-15.jpg">
```

In this code, the image tag tells the browser where to put the image on the page, while the src attribute gives the location to pull the image from. You can pull (use) images from online or from your computer. In the code above, we are pulling the image from online.

Do the following:

Find an image online that you would like to display in your HTML code and copy the image's URL.

You can do this as follows:

1. Search for the image you want on Google (such as "kitten").
2. Click on "Images" under the search bar.
3. Click on the image you want (it should then display on the right side on your screen).
4. Right click on the image and select "Copy image address." Paste this URL somewhere because we will be using it soon.

Open up your Test.html file (if not already open, one way to do this is to right click on your file on the desktop and select Edit with Notepad++) and replace the code with this:

```
<!DOCTYPE HTML>
<html>
    <body>
    <img src="insert the image's URL here"></img>
    </body>
</html>
```

Save and execute your code. Your image is displayed in the browser!

Now let's try displaying an image that's saved on your computer.

COMPUTER IMAGE

The route to a file is called a "file path". The "file path" is the address of a file and specifies the exact location of a file. It provides a "path" to the file.

File paths say things like, "You can find the file named 'Winter' inside the folder named 'Poems', which is inside the C drive." A drive is simply a location where data is stored. The C drive is where most files and software are stored on a computer.

The various components of a path are separated by a text character that is unlikely to be used in the name of a directory or a file. Usually this character is a slash, backslash or colon ("/", "\" or ":" respectively).

The "\" (backslash) symbol separates the different parts of a collection of electronic documents in computers and it has been used for a long time. It was created as an aid to organizing and finding the various files you might store on a computer.

In a file path, backslashes are used to show that one item is below another (in terms of hierarchy).

A hierarchy refers to arranging things according to rank or status. It refers to arranging items according to their relative importance or characteristics. Storage hierarchy refers to a system where various data storage devices are given a hierarchical importance ranking as far as how they are used. The primary factor influencing a given device's ranking is its response time – how long the device takes to return a requested piece of stored data when the computer requests the data. Faster response times are ranked higher.

In a file path, the item on the left is "above" the one on the right. If we take our earlier example, the file path would be written as:

```
C:\Poems\Winter
```

To prepare for the next assignment, download an image of your choosing (ensure to save it in your Downloads folder).

Written documents, pictures and videos are examples of different types of file formats. The data in these files is organized in a specific way, usually based on how the data will be used by various processes the computer does.

For example, the data in a file that is a written document will be organized in a very different way than the data in a file that is a video.

A common file format is "PDF". This stands for "Portable Document Format". This is a file format developed by the software company Adobe in the 1990s. This file format was developed to ensure consistent display of files regardless of what computer they're being displayed on.

File formats are indicated by their extension. The extension is a code that states what the file format is (text, image, video, etc.). Extensions are tacked on at the end of the file's name.

Let's say you have a file entitled "Contract" and it's stored as a PDF; the file would be named "Contract.pdf" – .pdf is the extension.

Another common file format and extension is ".txt" (short for "text file"). This is a file type that is pure, unformatted (no special design; plain) text. If you've ever used Notepad, it saves and runs .txt files.

As we covered earlier, your HTML files must always end with the .html extension.

If we want to display an image that's saved on your computer in our HTML code, we include the file path, as covered in the code below. The file path should be: `C:\Users\user\Downloads\File_Name.Extension`

But to ensure you have the correct file path, you can right click the image and select Properties – you'll see the file path to the image next to "Location:"

For example, if we had a picture named "Kitten" stored in our Downloads folder, this could be the file path: `C:\Users\user\Downloads\Kitten.jpeg`

Jpeg is a type of image and we must include the extension for the image in our file path. Otherwise, the image won't display.

Open up your Test.html file and replace the code with this:

```
<!DOCTYPE HTML>
<html>
    <body>
    <img src="C:\Users\user\Downloads\File_Name.Extension"></img>
    </body>
</html>
```

Save and run your code.

If it doesn't display correctly, check the following:

1. Did you save your image in your Downloads folder? If not, move it there.
2. Did you include the image file extension? If not, add it.
3. Did you enter the correct file path? If not, verify it and add it.

"HELLO, WORLD" IN JAVASCRIPT

Now, there are two ways we can add JavaScript to our HTML code. The first way is to just write the code inside the HTML using the `<script>` element.

`<script>` means we are inserting a script, and this is where we write our JavaScript code.

In coding, an "object" is something that has state and behavior. For example, a "dog" object could have "awake" as its state and "barking" as its behavior.

A "method" is something an object can do. It's a named sequence of events. A method should always have a meaningful name, such as CalculateIncomeTax – not Method_1 or something. That way you can get an idea of what it's supposed to do from reading the name.

"Alert" is a JavaScript method that displays an alert box (small window with an "OK" button).

Characters are letters, numbers and symbols – such as those found on a keyboard. As we mentioned earlier, strings (series of characters) are one of the six types of value. In JavaScript, we can create strings with double quotes (") or single quotes ('), such as: "Mary had a little lamb," or 'Its fleece was white as snow.' In our above code, "Hello, World!" is a string. It doesn't matter whether you use double or single quotes; they both perform the same functions.

Let's use JavaScript within our HTML file to display Hello, world! – edit your code as follows:

```
<!DOCTYPE HTML>
<html>
    <body>
        <script>
        alert("Hello, World!");
        </script>
    </body>
</html>
```

Save and execute this code. As a note: When we say "execute" here, we mean "Run the code in Google Chrome by clicking 'Run,' then clicking 'Launch in Chrome'." *You should always save your code before running it.*

Good job! We just made a pop-up!

When we include our JavaScript code inside of our HTML file, it's called inline JavaScript.

There's another way we can use JavaScript with HTML and that's by connecting our code to an external file. Meaning, we have an HTML file with your HTML code in it, and a *separate* file containing your JavaScript code.

Here's how that would work. First, create a new file and write this code:

```
alert("Hello, World!");
```

That is our JavaScript code.

Save this file on your desktop. Save it as "Test_1," as a "JavaScript file" (i.e., click File, Save As, write "Test_1" as the File name, and then click on JavaScript file under "Save as type").

To link our HTML file with our JavaScript file, we use the script element with the src attribute. Src is short for "source" – meaning, we are designating the source (location/name) of the file we are linking to.

Now, go back to your original HTML file and write this code:

```
<!DOCTYPE HTML>
<html>
    <body>
        <script src="Test_1.js">
        </script>
    </body>
</html>
```

Save and execute the code. We displayed the same alert box in a different way!

Keeping your JavaScript code in a separate file from your HTML code is the recommended way and considered best practice. It keeps your code better organized.

As a recap, we can execute our JavaScript in two ways:

1) Inline JavaScript (write all your Javascript within your HTML file), or

2) Write your JavaScript code in a separate, external file and referencing it (referring to it) in your HTML file.

While the latter option is the way to go when doing professional software development, we will use both of these methods throughout this book.

WINDOW.ALERT()

We can cause an alert window to pop up using the window.alert() method. In JavaScript, statements are the lists of instructions to be executed by the computer that make up a program. Statements are separated using a semicolon. It tells the computer, "Here is the end of this statement."

Delete the code in your HTML file and then write, save and execute this code:

```
<!DOCTYPE HTML>
<html>
    <body>
        <script>
        window.alert("Hello, world!");
        </script>
    </body>
</html>
```

Good job! You created a popup window with inline JavaScript.

DOCUMENT.WRITE()

You can also display text in JavaScript using the document.write() method. Delete the code in your HTML file and then write, save and execute this code:

```
<!DOCTYPE HTML>
<html>
    <body>
        <script>
        document.write("Hello, world!");
        </script>
    </body>
</html>
```

As you can see, there is more than one way to perform actions in programming – problems have multiple solutions!

END OF CHAPTER CHALLENGE

Write a program that includes the following:

- Create a webpage that has text and an image and successfully run it in the browser.
- Use an external JavaScript file (i.e., use <script src ="">) and document.write() to display text of your choosing.

CHAPTER FOURTEEN
VARIABLES AND STRINGS

A common action in coding is to assign variables. Variables are key in every programming language. We defined the term earlier, and to put the definition simply: it's a value that can be altered, depending on conditions or data passed to the program.

In mathematics, an operator is a symbol used to carry out a computation (that act of figuring out the amount of something using math). There are several different kinds of operators. Arithmetic operators, such as:
- + (add),
- - (subtract),
- / (divide), and
- * (multiply)

are used to perform math operations.

ASSIGNMENT OPERATOR

In JavaScript, the assignment operator is "="; it is used to assign value to a variable.

For example, to say that X = 10 in JavaScript, we are assigning the variable "X" the value "10."

Let's assign a variable in JavaScript. Delete your old code and write the following:

```
<!DOCTYPE HTML>
<html>
    <body>
        <script>
        var X = "Hello, world!";
        document.write(X);
        </script>
    </body>
</html>
```

Save and execute this code. We assigned variable X to the string value "Hello, world!"

CREATING A STRING

There are some characters and functions that cannot easily be put within quotes when writing strings in JavaScript.

For example, what if you wanted to display: Sally said, "I don't want to go"? You not only have additional quotation marks around Sally's statement, but you also have the single quote (apostrophe) within "don't."

The solution to this problem is to add a backslash (\) – this is referred to as "escaping" the character and tells the computer that the character that follows has a special meaning. For example: \' or \" means that the quote mark(s) following the backslash will not end the string but instead will be printed on the screen.

Also, if you want to start a new line, you can use the HTML <br> tab, which is like pressing Enter. This means "line break." A line break literally creates a new line of text – again, like pressing Return/Enter on your keyboard.

Let's try this out. Delete your previous HTML/JavaScript code and write this code:

```
<!DOCTYPE HTML>
<html>
<body>
    <script>
    document.write("Lisa told Bart, \"Stop it, Bart! Or I'll tell dad!\"<br>\"Eat my shorts!\" Bart responded.");
    </script>
</body>
</html>
```

Save and execute your code. We have successfully displayed quotation marks and an apostrophe!

Wait a second... What do you do if you need to display a backslash in your code? Simple, write two of them!: \\

CONCATENATING A STRING

Concatenate means to connect items together, like links in a chain. It means to take one piece of data and stick it on the end of another piece of data. For example, concatenating the string "dev" and the string "ices" makes the text "devices."

To concatenate a string in JavaScript, you use the + operator.

Delete your JavaScript code and write the following:

```
<!DOCTYPE HTML>
<html>
<body>
    <script>
    document.write("\"Be who you are and say what you feel,"
    + "because those who mind don\'t matter and those who matter don\'t mind.\"
    + "-Dr. Seuss");
    </script>
</body>
</html>
```

Save and execute your code. Something is not quite right... We are missing a space between "feel," and "because."

To fix this, simply add a space before "because" as follows:

```
" because those who mind don\'t matter and those who matter don\'t mind.\""
```

Save and execute your code. Now it displays correctly!

MULTIPLE VARIABLES

You can also assign multiple variables in one statement. Write this code:

```
<!DOCTYPE HTML>
<html>
<body>
    <script>
        var Family = "The Arezzinis", Dad = "Jeremiah",
            Mom = "Hermoine", Daughter = "Penny", Son = "Zorro";
        document.write(Dad);
    </script>
</body>
</html>
```

Save and run your code. You assigned multiple variables and then displayed one.

STR.FONTCOLOR(""):

A cool thing we can do with our strings is change the font color using the `str.fontcolor` method.

Open a Notepad++ document and write this code:

```html
<!DOCTYPE HTML>
<html>
<body>
    <script>
        var blues = "I have the blues.";
        var blues = blues.fontcolor("blue");
        document.write(blues);
    </script>
</body>
</html>
```

Save and execute your code. We created a variable that was a string called "blues." We assigned that variable the color blue and then displayed it. See how that works?

END OF CHAPTER CHALLENGE

Write a program that includes the following:

- Assign multiple variables and display one utilizing the document.write() method, and
- Create a string, concatenate a string and change the string's font color.

CHAPTER FIFTEEN
EXPRESSIONS AND STATEMENTS

Expressions are numbers, symbols, and operators grouped together that show the amount of something. Basically, an expression is a written math problem.

In computers, an expression is a combination of values that are computed by the computer. There are different ways to write out expressions, depending on which language you are programming in.

For example, `Name = "What person types in the name box"` could be an expression. Also, `x + 5` is an expression.

As a reminder, statements are computer instructions. These are the instructions that are used by people as they create computer programs. The simplest of these might be things like "print," "delete," "add," "multiply," etc.

For example, the "print" statement tells your computer to print whatever text you typed as part of the command.

In JavaScript, expressions and statements are different things. An expression results in a value, while a statement performs a task – a program is basically a series of statements.

There are basically two types of expressions:

1) Ones that have a value (or result in a value), and

2) Those that assign a value to a variable.

The following JavaScript code is an expression:

```
3 + 3
```

Now, consider the following:

```
Document.write(3 + 3);
```

This is a statement. Within this statement, the expression `3 + 3` is contained.

Let's write a statement. Open your HTML file and write this code:

```
<!DOCTYPE HTML>
<html>
<body>
    <script>
    document.write(1 + 1);
    </script>
</body>
</html>
```

Save and execute your code. You just performed math with JavaScript and displayed the result!

END OF CHAPTER CHALLENGE

Complete the following:

- Write your own statement and ensure it displays a result.

CHAPTER SIXTEEN
FUNCTIONS

A JavaScript function is a repeatable block of code that executes certain actions and performs tasks. You execute a function by calling it. This is also called "invoking" the function.

"Invoke" literally means to request or ask for something. In coding, it is to cause something to be carried out (performed).

In JavaScript, a keyword identifies actions to be performed. There is a *function* keyword.

Every spoken language has a general set of rules for how words and sentences should be structured. These rules are known as the syntax of that particular language. In programming languages, syntax serves the same purpose. Syntax is the rules you must follow when talking to a computer and telling it what to do. There are many languages you can use to program a computer. Each language has its own syntax. Failing to use the syntax of a particular language correctly can mean that whatever you are designing will not work at all.

Syntax is the arrangement of words and phrases to create well-formed sentences in a language. In computer science, it is the language that allows the user to write out the program and the rules about how to write code properly.

The syntax for writing a JavaScript function is: the keyword, then the name, then parentheses.

Functions are valuable because of code reusability – you can invoke functions over and over. Since functions in JavaScript contain properties (state) and methods (behavior), they are basically objects.

Functions are useful so you don't have to keep retyping code; you can instead simply invoke the function whenever needed.

We will write a function shortly, but first...

<p> ELEMENT

The `<p>` element is used to create paragraphs.

Open your HTML file and write this code:

```
<!DOCTYPE HTML>
<html>
    <body>
        <p>This is a paragraph!</p>
    </body>
</html>
```

Save and execute your code. Good job! You created a paragraph.

IDs

"Id" is short for "identification." One of the most popular attributes is the id attribute.

The id attribute specifies a unique name for an HTML element.

For example, you can assign a specific <p> element the id (name) "paragraph" as follows:

```
<p id="paragraph">This is some text.</p>
```

Now that you have that id, you can reference (bring up; refer to) it in your code later.

You will learn exactly how to do this and why it's useful shortly.

Ids are specific and you can only utilize one id per element. For example, you cannot assign the same <h1> element two different ids. As a note, the <h1>-<h6> elements create headers (section heads) in HTML. <h1> is the largest font size, and it gets smaller with each higher number (<h2>, <h3>, etc.). Also, you cannot use the same id name for two different elements.

For example, this would be incorrect:

```
<h2 id="heading"></h2>
<h3 id="heading"></h3>
```

Ids must be specific to one element and cannot contain any spaces.

DOCUMENT.GETELEMENTBYID

The `document.getElementById` method returns an element.

The element has an id attribute with a specific value assigned to it. It is used mainly to control or get information from an element within your code. If it can't find the element with the specified value, it will return "null." Null means "having a value of zero; no value."

You'll understand it better once we put it into action!

Using the `<button>` element, we can create a button like this:

```
Button
```

Now let's create a function and use the `document.getElementById` method. First, open up your HTML code in Notepad++ (Test.html) and write this code:

```html
<!DOCTYPE HTML>
<html>
<body>
    <button onclick="My_First_Function()">
        Click me!
    </button>
    <p id="Irish"></p>
<script>
    function My_First_Function() {
        var String = "Kiss me, I'm Irish!";
        var result = String.fontcolor("green");
        document.getElementById("Irish").innerHTML =
        result;
    }
</script>
</body>
</html>
```

Save and execute your code. You created a function and a functional button!

Read through your code and try to figure out what we just did.

Everything inside the curly brackets { } is our function.

We used the ID attribute and assigned the button element the value "Irish."

Then we returned the button element by calling the "Irish" value that we assigned earlier (when writing the ID attribute).

+= OPERATOR

You can also use the += operator to concatenate a string. Write this code:

```html
<!DOCTYPE HTML>
<html>
    <body>
        <p id="Concatenate"></p>
        <script>
        Sentence = "I am learning ";
        Sentence += "a lot from this book!";
        document.getElementById("Concatenate").innerHTML = Sentence;
        </script>
    </body>
</html>
```

Save and execute your code. Well done!You concatenated a string using the += operator.

END OF CHAPTER CHALLENGE

Write a program that includes the following:

- A function, and
- The document.getElementById() method.

CHAPTER SEVENTEEN
NUMBERS

For simplicity, we are going to keep our HTML and JavaScript files separate in this chapter. So, first, open up your HTML file in Notepad++ (Test.html) and ensure the code states the following:

```
<!DOCTYPE HTML>
<html>
    <body>
        <p id="Math"></p>
        <script src="Test_1.js">
        </script>
    </body>
</html>
```

Save your code. For the next few steps, we will refer to this file as your "HTML code."

Binary is a form of counting and performing math that uses only two numbers. The word "binary" comes from the Latin word *binarius*, meaning "two together" or a "pair." All quantities in binary are represented by numbers that use a 1 and/or a 0. In fact, any number can be written in binary.

The ones and zeros in a computer are called bits (short for "binary digits"). Bits are held in the computer by using very small devices, each of which can change to represent either a 1 or a 0.

As we discussed earlier, another JavaScript value type is "numbers." To store numbers, JavaScript uses 64 bits. 64 bits allows for 18,000,000,000,000,000,000 different numbers.

Arithmetic is the most common use of numbers in JavaScript. We already defined "operator," but as a reminder, an operator is a symbol used to carry out a computation. Arithmetic operators include +, -, /, and *. In JavaScript, we can use these to perform arithmetic.

Now, open up your JavaScript file (Test_1.js – referred to as your "JavaScript code" for the next few steps) and write this code:

```
var Simple_Math = 2 + 2;
document.getElementById("Math").innerHTML = "2 + 2 = " + Simple_Math;
```

Save your code. Now, if you run your JavaScript file, it won't work! That's because you need to run your HTML file that is linked to the JavaScript code. Execute your HTML file.

SUBTRACTION

Let's do some subtraction – edit your JavaScript code as follows:

```
var Simple_Math = 5 - 2;
document.getElementById("Math").innerHTML = "5 - 2 = " + Simple_Math;
```

Save your JavaScript code and execute the HTML file.

MULTIPLICATION

To multiply, edit your JavaScript code as follows:

```
var Simple_Math = 6 * 8;
document.getElementById("Math").innerHTML = "6 X 8 = " + Simple_Math;
```

Save your JavaScript code and execute the HTML file.

DIVISION

We can divide numbers – edit your JavaScript code as follows:

```
var Simple_Math = 48 / 6;
document.getElementById("Math").innerHTML = "48 / 6 = " + Simple_Math;
```

Save your JavaScript code and execute the HTML file.

MULTIPLE ARITHMETIC OPERATORS

Let's say you would like to multiply, subtract, add *and* divide numbers all at once. Edit your JavaScript code as follows (this is *one* way to do it):

```
var Simple_Math = (1 + 2) * 10 / 2 - 5;
document.getElementById("Math").innerHTML = "1 plus 2, multiplied by 10, " +
    "divided in half and then subtracted by 5 equals " + Simple_Math;
```

Save your JavaScript code and execute the HTML file.

MODULUS OPERATION

The modulus operation finds the remainder after dividing two numbers. Remember back in math class? In division, the remainder is the amount left over after dividing two numbers. For example: 5 ÷ 2 would have a remainder of one.

A couple other math terms are "dividend" (the number being divided) and "divisor" (the amount the number is divided by). In our earlier example, 5 is the dividend and 2 is the divisor.

"Modulus" comes from the Latin word *modulus*, meaning "measure."

The modulus is the remainder after the dividend has been divided by the divisor.

In JavaScript, the % symbol represents the modulus operator – which could be called the remainder operator. Let's put this to use. Open up your JavaScript code and write the following:

```
var Simple_Math = 25 % 6;
document.getElementById("Math").innerHTML=
    "When you divide 25 by 6 you have a remainder of: "
    + Simple_Math;
```

Save your JavaScript code and execute the HTML file.

UNARY AND BINARY OPERATORS

"Unary" means having, made up of, or acting on one component, item or element. An "operand" is the number that is being dealt with in a mathematical operation. It is not the action being taken with the number – it is the number itself.

For example, in 5+6, the operands are 5 and 6. The "+" is the operator.

A "unary operator" is an operation that only contains a single operand. An example of this would be + 5.

A "binary operator" is an operation that requires two operands – one operand before the operator and one after. An example of this would be 5 + 5.

There is a unary operator called a "negation operator." Negate literally means to deny or contradict something. The word "negation" in "negation operator" means "negative; not positive."

This returns the opposite or negative form of something.

Let's write this code out in a new text document:

```
<!DOCTYPE html>
<html>
    <body>
        <p id="Negation"></p>
        <script>
        var x = 10;
        document.getElementById("Negation").innerHTML = -x;
        </script>
    </body>
</html>
```

Save and execute your code. Note: the syntax for the negation operator is "-x."

INCREMENT OPERATOR

An increment is an addition or increase to something. "To increment" means to increase. The increment operator in JavaScript is written as ++ and counts one step up.

Write this code:

```
<!DOCTYPE html>
<html>
    <body>
        <script>
        var X = 5;
        X++;
        document.write(X);
        </script>
    </body>
</html>
```

Save and execute your code – it incremented from 5 to 6.

The opposite of this is the decrement operator. Decrement basically means to count down.

Write this code:

```
<!DOCTYPE html>
<html>
    <body>
        <script>
        var X = 5.25;
        X--;
        document.write(X);
        </script>
    </body>
</html>
```

This should show up as "4.25."

MATH.RANDOM()

To return a random number between 0 and 1, write this code:

```
<!DOCTYPE html>
<html>
    <body>
        <script>
        window.alert(Math.random());
        </script>
    </body>
</html>
```

Save and run your code. Good job!

To have a random number displayed between ____ and ____ (such as between 0 and 100), write this code:

```
<!DOCTYPE html>
<html>
    <body>
        <script>
        window.alert(Math.random() * 100);
        </script>
    </body>
</html>
```

Save and run your code. Excellent!

END OF CHAPTER CHALLENGE

Write a program that includes the following:

- A binary operator,
- A unary operator,
- The modulus operator,
- Math.random(), and
- The increment or decrement operator.

NOTE: Ensure that each of the above displays (prints) the result.

CHAPTER EIGHTEEN
JAVASCRIPT OBJECTS

We talked about objects earlier. Objects are items that can be represented in a computer program. They are often meant to represent real-world things.

You are surrounded by objects – your dog, the TV, etc. Objects have state and behavior. The state of an object would be its size, color, etc. The behavior of an object would be what the object does – the actions it takes.

A race car could be an object. Characteristics that describe its state could be: engine type, engine horsepower, wheel size, gas tank capacity, etc. Characteristics that describe its behavior could be: accelerate, decelerate, turn right, turn left, etc. In computers, objects are parts of computer programs and share a similarity to real life objects: they have a state and behavior.

Let's look at a further example: an application might work with a "Customer" object. Again, the state of an object would be the characteristics and attributes of the object. The behavior would be what the object could do. In our example, the "Customer" object could have states like "active" or "deleted." It could have behavior like "Upgrade Rewards Level" or "Add to Family Account."

Objects are one of the first subjects you think about when designing a program. An object is something on a computer that you can click on, interact with, move around, etc. It can also be something behind the scenes that is made up of data and procedures to manipulate data. An object is what actually runs in the computer.

As another example, let's say you wanted to make two different types of cars on a computer; each car would be an object. Each object (car) would have its own size, shape, color, speed, distance it could travel without needing more gas, etc.

DICTIONARIES

In computer programming, dictionaries are a specialized type of list. The first item is the key, and the second item is the value. Together, they are a "key-value pair."

Key-value pair is abbreviated "KVP." A KVP is a set of two linked data items that consists of: a key (a unique identifier for some item of data), and the value (the data that is identified by the key). The key is the unique name, and the value is the content.

Collections of KVPs (i.e., dictionaries) are often used in computer programs.

Below is an example of a collection of KVPs that might be used in a computer program for a school. Here, the KEY is used to store the name of a course at the school, and the VALUE is the description of the course – this is a dictionary:

KEY	VALUE
SOUND	Boom!
SEASON	Summer
WEATHER	Stormy
DAY	Saturday
MONTH	August

Note that in this list, you could not have a second Key-Value Pair that used the Key "SEASON" or "DAY" because the keys in a given collection of KVPs must be unique.

A "dictionary" is "a key-value pair collection."

In a dictionary, multiple values can be assigned to one key. This can be useful because you can look up various values by their key. In the below example, each key has three values assigned to it:

KEY	VALUES
0	Eddard Stark | 35 years old | Male
1	Cersei Lannister | 32 years old | Female
2	Jon Snow | 18 years old | Male
3	Daenerys Targaryen | 20 years old | Female

To create a dictionary in JavaScript, write this code in your HTML file (Test.html):

```html
<!DOCTYPE html>
<html>
    <body>
        <p id="Display"></p>
        <script>
        var Animal = {
            Species:"Dog",
            Color: "Black",
            Breed: "Labrador",
            Age:5,
            Sound:"Bark!"
        };
        document.getElementById("Display").innerHTML = Animal.Sound;
        </script>
    </body>
</html>
```

Save and execute your code. The output is: Bark! because this was associated with the key "Sound."

NOTE: Due to the fact that 5 is the value type "number," we don't have to place it within quotation marks.

ASSIGNING MULTIPLE VALUES

We can assign multiple values to the keys in our dictionary. Edit your previous code to the following:

```html
<!DOCTYPE HTML>
<html>
    <body>
        <p id="Display"></p>
        <script>
            var Food = {
                Vegetables: ["broccoli", "carrot", "lettuce"],
                Meat: ["hotdog", "steak", "hamburger"],
                Fruit: ["strawberry", "orange", "apple"],
                Quantity: [10, 21, 182]
            };
            document.getElementById("Display").innerHTML = Food.Fruit;
        </script>
    </body>
</html>
```

OPERATORS AS WORDS

Most operators are written as symbols: +, -, /, etc. But some are words. An example of this is the delete operator.

Edit your previous code to the following:

```
<!DOCTYPE HTML>
<html>
    <body>
        <p id="Delete"></p>
        <script>
            var Animal = {
                Species: "Dog",
                Color: "Black",
                Breed: "Labrador",
                Age: 5,
                Sound: "Bark!"
            };
            delete Animal.Sound;
            document.getElementById("Delete").innerHTML = Animal.Sound;
        </script>
    </body>
</html>
```

Save and execute your code. We deleted the value associated with the key Animal, so "undefined" is returned because the value is no longer defined in our dictionary.

TYPEOF OPERATOR

This operator gives the value type of a variable. Write this code:

```
<!DOCTYPE html>
<html>
    <body>
        <script>
        document.write(typeof "This is a string");
        </script>
    </body>
</html>
```

Save and execute your code. It returned "string."

Now change your typeof operator to this:

```
document.write(typeof 247365);
```

Save and execute your code. It returned "number" – another JavaScript value.

END OF CHAPTER CHALLENGE

Write a program that includes the following:

- A dictionary.
- Delete a KVP pair from your dictionary using the delete operator.
- Use the typeof operator.
- Try to create a dictionary with two identical keys, then attempt to display the keys using the document.getElementById() method and see what happens.

CHAPTER NINETEEN
FLOATING POINT

It's time to get mathematical!

Factors are numbers that can multiply together to get another number. For example, 4 and 5 are factors of 20 because 4 × 5 = 20

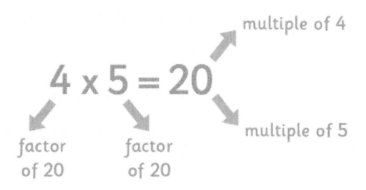

Superscripts are smaller characters typically displayed toward the upper right of another number - like this: 4^{12}

Superscripts usually indicate "power." Power refers to how many times a number is multiplied by itself. For example, 10^4 means 10 to the fourth power – which is 10,000 (10 multiplied by 10 multiplied by 10 multiplied by 10).

A notation is a symbol or symbols used to represent numbers or amounts – such as an E placed at the end of a number. This notation means "10 to a given exponent." An "exponent" is the power to which a number is multiplied. In the 10^4 example given above, 4 is the exponent.

A floating point is a specific way of representing large numbers. Instead of writing out the entire number, you can write the first part of it and then a factor to multiply it by. This makes it so that by performing the multiplication, you get essentially the same number you would have by writing it out in full. When you do this, you move the decimal point in the number; hence the name "floating point."

For example, take the number three thousand, four hundred twenty-five. Written normally, it would be: 3425 – written in floating point notation, it could be: 3.425×10^3. Since 10^3 is 1000, this means "3.425 times 1000" or 3425.

A "float" is a type of data used by computers. The "float" type is used to store numbers with a very high degree of accuracy. It is primarily used for mathematical purposes.

Let's say we want a shorter way to write 100,000. We could write 10E 4.

10E 4 means 10×10^4 (10 multiplied by 10 four times).

In JavaScript, all numbers are floating point numbers, and the limit of floating point numbers is 1.797693134862315E+308 (meaning that JavaScript can not handle a larger floating point number). Any number that is larger than 1.797693134862315E+308 is considered "infinity." As a note, this is a massive number far beyond the capability of most calculators and would take an excessive amount of lines to display.

OTHER NUMBERS

In JavaScript, there are three unique values that are considered to be numbers but actually do not act in the same way. They are:

- NaN (short for "Not a Number")
- Infinity (positive infinity)
- -Infinity (negative infinity)

The best way to understand these is to put them into action. Here's an example: in a *new* Notepad++ document, write this code:

```
<!DOCTYPE html>
<html>
    <body>
        <p id="Not_a_Number"></p>
        <script>
            document.getElementById("Not_a_Number").innerHTML = 0/0;
        </script>
    </body>
</html>
```

Save your file as an .html document with a name of your choosing. Then execute your code. The result is NaN because you can't divide 0 by 0.

We can check whether or not something is a number by using the isNaN() function. Edit your code to the following:

```
<!DOCTYPE html>
<html>
    <body>
        <p id="Not_a_Number"></p>
        <script>
        document.getElementById("Not_a_Number").innerHTML = isNaN('Hello World');
        </script>
    </body>
</html>
```

Save and run your code. As you can see, the result is "true" because "Hello World" is *not* a number. Let's change our code to the following:

```
<!DOCTYPE html>
<html>
    <body>
        <p id="Not_a_Number"></p>
        <script>
        document.getElementById("Not_a_Number").innerHTML = isNaN('007');
        </script>
    </body>
</html>
```

Note: the only difference was that we changed "Hello World" to "007." Save and run your code. Now it is false because 007 *is* a number.

Now, delete your code and write the following:

```
<!DOCTYPE HTML>
<html>
    <body>
    <p id="Infinity"></p>
        <script>
        document.getElementById("Infinity").innerHTML =
        1.7976931348623157E+309;
        </script>
    </body>
</html>
```

Save and execute your code. You can see the result "infinity." This is because we entered a number larger than 1.797693134862315E+308.

To see -Infinity, simply change your code to a number lower than -1.797693134862315E+308, such as the number below:

```
<!DOCTYPE HTML>
<html>
    <body>
    <p id="Infinity"></p>
        <script>
        document.getElementById("Infinity").innerHTML =
        -1.7976931348623157E+309;
        </script>
    </body>
</html>
```

END OF CHAPTER CHALLENGE

Create your own program that returns the following:

- Infinity,
- -Infinity, and
- NaN.

CHAPTER TWENTY
BOOLEAN LOGIC

"Logic" refers to actions, behavior and thinking that makes sense. When speaking about computers, logic is the rules that form the foundation for a computer in performing certain tasks. Computer logic is the guidelines the computer uses when making decisions.

Logical operators like "and," "or," and "not" are used to evaluate whether an expression is true or false.

George Boole was an English mathematician that developed Boolean logic. "Boolean logic" is a form of logic in which the only possible results of a decision are "true" and "false". There aren't any vague or "almost" answers to a calculation or decision – black or white, no gray.

An example of Boolean logic would be answering questions with only "yes" or "no".

Boolean logic is especially important for the construction and operation of digital computers because it is relatively easy to create a machine where the result of an operation is either "true" or "false."

This is done by comparing two or more items – items that can only be "true" or "false."

Some common examples of such Boolean comparisons are "AND" and "OR".

With the Boolean comparison called "AND", the comparison is true *only if all of the involved comparisons are true.*

Let's look at some examples to show how this works:

In the following AND comparison, the result is true:

```
5 is more than 3 AND 10 is more than 5
```

Let's break it down.

There are three comparisons happening here:

1. Comparing 5 and 3 to see if 5 is larger than 3 (is 5 larger than 3?)

2. Comparing 10 and 5 to see if 10 is larger than 5 (is 10 larger than 5?)

3. Comparing the results of those two comparisons, using the Boolean comparison "AND" (are both comparisons true?)

This is the overall comparison.

It is true that 5 is greater than 3, so the first comparison is true.

It is also true that 10 is greater than 5 – so the second comparison is true as well.

A Boolean AND comparison is true if the other comparisons are all true – so in this example, the overall comparison is true, since the first comparison is true *and* the second comparison is true.

In this next example, the result is false (not true):

```
5 is more than 7 AND 10 is more than 5
```

Even though 10 is more than 5, 5 is not more than 7 – so the overall comparison is not true.

A "condition" is an item that must be true before something else occurs. In the AND comparison above, these are the two conditions checked for:

1. 5 is more than 7

2. 10 is more than 5

They're conditions because the outcome is conditional upon (dependant on) these two comparisons.

A Boolean OR comparison checks for whether one *or* both conditions are true. Here is an example:

```
4 is less than 9 OR 8 is less than 7
```

The result would be true because at least one of the comparisons is true (4 is a smaller number than 9).

In the following example, the result would be false since neither is true:

```
8 is less than 4 OR 9 is less than 3
```

And in this example, the result would be true because one or both (in this case both) are true:

```
7 is less than 8 OR 2 is less than 5
```

As a reminder, in writing instructions for a computer, we would use the greater and lesser symbols (> and < respectively). For example: 7 > 3 means "seven is greater than three", and 4 < 9 means "four is less than nine".

So, for example, instead of 10 is greater than 2, we would write:

```
10 > 2
```

And, again, if we wanted to say "greater than or equal to", we could use the symbol >=.

For example:

```
10 >= 8
```

This would be true.

Like virtually all other programming languages, the Boolean values in JavaScript are true and false.

Delete the last code you wrote and write the following:

```
<!DOCTYPE html>
<html>
    <body>
        <script>
        document.write(10 > 5);
        </script>
    </body>
</html>
```

Save and execute this code. You should see "true" displayed. That is because it is true that 10 is larger than 5. If we change our code to the following, we will see "false":

```
<!DOCTYPE html>
<html>
    <body>
        <script>
        document.write(10 < 5);
        </script>
    </body>
</html>
```

This is Boolean logic.

CONSOLE.LOG METHOD

The `console.log` method can be used to display data in the console within the browser. The console is a tool you can use within your browser to debug code.

You can access the console in a few ways, including:

a. Pressing the F12 key and then clicking on "Console," or

b. Clicking the three upright dots in the top-right of the browser, selecting "More tools," clicking "Developer tools," and then selecting "Console."

Replace the code you last wrote with this:

```
<!DOCTYPE html>
<html>
    <body>
        <script>
        console.log(2 + 2);
        </script>
    </body>
</html>
```

Save and run your code. Don't see anything? That's because it's displayed in the console. Open the console, and you should see "4"!

TYPE COERCION

In English, "coercion" means to force or persuade someone, typically by use of threats or punishment. "Type coercion" in JavaScript is when the operands of an operator are different data types (such as "string" and "number" – like: "five" + 5).

The type coercion converts the value to the type it chooses. Let's try this out. Write this code:

```
<!DOCTYPE html>
<html>
    <body>
        <script>
        document.write("10" + 5);
        </script>
    </body>
</html>
```

Many other programming languages would return an error if we were to include two different data types within an operation.

END OF CHAPTER CHALLENGE

Write a program that includes the following:

* Boolean logic that returns "true,"
* Type coercion, and
* Display "false" in the console.log using Boolean logic.

CHAPTER TWENTY ONE

COMPARISON OPERATORS

In JavaScript, we can utilize double and triple equal signs.

== (double equal sign) is a symbol used to show that a comparison should be made. Specifically, this "==" symbol is an instruction to check whether the data on the left side of the symbol is equal to the data on the right side. The answer to this comparison is an answer of "true" or "false."

When the "==" symbol is used to check for equality, usually it is used like this: [first item to be compared] == [second item to be compared]

For example: (10 + 5) == 15

In this example, we are asking the computer to check whether the result of adding 5 and 10 is equal to 15.

When the "==" symbol is used in this manner, we are asking the computer to give a response after checking for equality. Usually, this result is given in a "true" or "false" manner. Here, the response would be "true," since (10 + 5) is equal to 15.

Another example: (10 + 6) == 15

Here, the computer would respond with "false," since (10 + 6) is not equal to 15.

=== (triple equal sign) is used to show that a comparison should be made. Specifically, this "===" symbol is an instruction to check whether the data on the left side of the symbol is equal to the data on the right side, and that it is the same type of data (data type) as that on the right. The answer to this comparison is an answer of "true" or "false."

For example, let's say you want to check whether two birth dates are equal. You have two pieces of data in the computer that represent these two birth dates:
1. "DateOfBirth1" is data of type "Date," and the value of the data is "1/1/1970."
2. "DateOfBirth2" is data of type "Date," and the value of the data is "1/1/1970."

You would use the "===" symbol like this:

```
DateOfBirth1 === DateOfBirth2
```

This tells the computer to check whether the two pieces of data are equal in both VALUE and TYPE. Since they are in this case, the computer responds with "true."

Like all code, it's best understood by writing it! Now, let's use a separate HTML document and JavaScript document again. First, open up an HTML file (if you don't have one, just make a new one) and write the following code:

```
<!DOCTYPE HTML>
<html>
    <body>
        <script src="Test_1.js">
        </script>
    </body>
</html>
```

Ensure this file is saved as HTML.html. Then open up your JavaScript code file (entitled Test_1.js – if you don't have this file, create a new file) and type the following code:

```
document.write(10 == 10);
```

Again, ensure this is saved as Test_1.js. Then run your HTML.html code. It should return "true" (because 10 is equal to 10). To make it return "false," simply change a number in your JavaScript, such as:

```
document.write(5 == 10);
```

Now, let's use ===. Write the following in your JavaScript code file:

```
X = 10
Y = 10
document.write(X === Y);
```

Save it and run the HTML code. This returns "true" because they are both variables and both have the value 10. You can make this false by changing it to:

```
X = 10
Y = 5
document.write(X === Y);
```

Save it and then run the HTML code – it returns "false." Even though they are both variables, the values are different. We can also return false by writing the following JavaScript code:

```
X = "Ten"
document.write(X === 10);
```

END OF CHAPTER CHALLENGE

Write a program that includes the following:

- Double equal signs (==) and type coercion (i.e., use a variable and a value) and get a "true" result displayed.
- Triple equal signs (===) to produce "true" and "false."

CHAPTER TWENTY TWO
LOGICAL OPERATORS

In JavaScript, there are three Boolean logical operators:

1. AND   written: **&&**
2. OR   written: **||**
3. NOT   written: **!**

The **&&** operator determines the logic between values or variables, such as verifying that _____ and _____ are true (both must be true to return "true." If one or both is false, the code will return "false").

Write this code:

```
<!DOCTYPE html>
<html>
    <body>
        <script>
        document.write(5 > 2 && 10 > 4);
        </script>
    </body>
</html>
```

Our code returns "true" because five is greater than two, and ten is greater than four. We can make our code return "false" by changing one or both of the comparisons, such as:

```
<!DOCTYPE html>
<html>
    <body>
        <script>
        document.write(5 > 10 && 10 > 4);
        </script>
    </body>
</html>
```

Write, save and execute this code and it will return false.

The || (or) operator means what it sounds like and is best described through utilization. Write this code:

```
<!DOCTYPE html>
<html>
    <body>
        <script>
        document.write(5 > 10 || 10 > 4);
        </script>
    </body>
</html>
```

Save and execute your code. It returned "true" because, while 5 is not greater than 10, 10 is greater than 4. We can make it return "false" with the following code since neither is true:

```
<!DOCTYPE html>
<html>
    <body>
        <script>
        document.write(5 > 10 || 10 > 20);
        </script>
    </body>
</html>
```

The ! (not) operator checks whether or not something is true. If _____ is not present, "true" will be returned.

Write this code:

```
<!DOCTYPE html>
<html>
    <body>
        <p id="Not"></p>
        <script>
        document.getElementById("Not").innerHTML = !(5 > 10);
        </script>
    </body>
</html>
```

Save and run your code. The result is "true" because 5 is *not* greater than 10.

And of course, if you want a "double negative," write this code:

```
<!DOCTYPE html>
<html>
    <body>
        <p id="Not"></p>
        <script>
        document.getElementById("Not").innerHTML = !!(20 > 10);
        </script>
    </body>
</html>
```

Save and execute your code. We are saying that 20 is *not not* greater than 10.

CHAPTER TWENTY THREE
TERNARY OPERATORS

Ternary means "made up of three parts." A ternary operator operates on three values. It can be used to assign a value to a variable, based on a condition. This is also referred to as a conditional operator in that it assigns a value to a variable based on a condition.

The syntax of this is:

```
Name_of_variable = (condition) ? Value_1:Value_2
```

The ternary operator is "?". Write this code:

```
<!DOCTYPE html>
<html>
    <body>
        <script>
        document.write(Bigger = (5 > 1) ? "The number on the left is bigger":
        "The number on the right is bigger");
        </script>
    </body>
</html>
```

Save and execute your code.

In this code, we said, "If it is true that 5 is bigger than 1, display 'The number on the left is bigger.'" If you change the numbers or flip the symbol to <, you can change the outcome of your code.

Input is a command that has the user input data. It allows users to type in information within your program.

Below is something more elaborate that we can do with ternary operators:.
Write this code:

```
<!DOCTYPE html>
<html>
    <body>
        <p>Riders must be at least 52 centimeters tall to ride.</p>
        <input id="Height" value="52" />
        <button onclick="Ride_Function()">Click here</button>
        <p id="Ride"></p>
        <script>
        function Ride_Function() {
            var Height, Can_ride;
            Height = document.getElementById("Height").value;
            Can_ride = (Height < 52) ? "You are too short":"You are tall enough";
            document.getElementById("Ride").innerHTML = Can_ride + " to ride.";
        }
        </script>
    </body>
</html>
```

Save and execute your code.

END OF CHAPTER CHALLENGE

Write a program that includes the following:

- A ternary operator, and
- The input command.

CHAPTER TWENTY FOUR
OBJECT-ORIENTED PROGRAMMING

As we covered earlier, objects are items within computer programs that have both state and behavior. The state of an object could be the weight, color, height, etc. While the behavior of an object would be what the object does – the actions it takes – such as jump, run, sit, etc.

Object-oriented is an approach to programming that focuses on objects and data (as opposed to consecutive actions or some other approach).

You create objects by creating what are called "classes," using one of the many specialized computer programming languages. Classes are used to describe one or more objects. This would be created by creating (or "declaring") a class called a "Customer" class. It is important to know that when you first create this class, you are describing the POTENTIAL characteristics and behavior of that TYPE of thing. You will still need to create an actual one of those things. That process is called "creating an INSTANCE" of the class, where "instance" means "an actual one" of the things described when you declared the class. When you do this, the data that makes up the object is kept in the computer's memory.

For example, in a computer program designed to track pay records for all employees, you could have a class called "Employee." Each program element representing an actual employee (John Smith, Sally Jackson, etc.) would be an instance of the "Employee" class. You would have one instance for each employee you entered into the computer. Each time you created an "employee" object, the computer would first find the "employee" class, then use the definition of that class in creating that particular "employee" instance using the data for the actual employee.

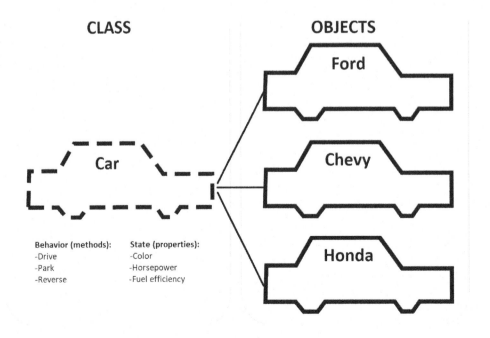

CONSTRUCTORS

A "constructor" is a special part of a class (a template for defining an object [item with state and behavior]).

The constructor is a special part of the class that describes the default state of any new instance of the class that gets created. In other words, it gives default values for the properties of the class.

For example, let's say you had a class called "Customer." The computer code to create the "customer" class might look something like this:

```
class Customer
{
    string FullName;
    Boolean Active;
}
```

Now let's say that whenever you created a new instance of the "customer" class, you wanted it to be an active customer – in other words, you wanted the property "Active" to be set to "true".

To do this, you would make a constructor for the "customer" class. It would be a small subprogram, inside the class, that would be used every time an instance of the "customer" class was created.

The constructor might look something like this:

```
Customer(string name) {
    Fullname = name;
    Active = true;
}
```

The entire class would look like this:

```
class Customer
    {
    string FullName;
    Boolean Active;
    Customer(string name){
        Fullname = name;
        Active = true;
    }
}
```

The constructor would be used by asking for an instance of the "customer" class to be created and passing along the desired name of the customer.

The call to create the instance of the class would look like this:

```
Customer cust = new Customer("Brenda Smith");
```

This creates a new instance of the "customer" class using the constructor inside the class. The constructor uses the string "Brenda Smith" that it was given to set the value of the property "FullName" and uses the instruction in the constructor to set the property "Active" to "true."

The new instance of the "customer" class, which is the variable "cust" here, will, therefore, have the properties:

```
FullName = "Brenda Smith";
Active = true;
```

Note: The above code is given as an example and is not meant to be executed. We will create a constructor soon.

KEYWORDS

As discussed earlier, JavaScript keywords identify actions to be performed. For example, var is a keyword we use to assign variables.

An important aspect to keywords is that they cannot be used to name variables or functions. For example: you can't say var var = "variable".

Another type of keyword is the "new" keyword, which is used to create new objects.

We also have the "this" keyword. The "this" keyword indicates the object that is the owner of the code. When used within an object, the value of "this" is the object. When used with a function, the value of "this" is the object that possesses the function.

Now we just covered a lot of information about constructors, the "new" keyword, the "this" keyword", and more. Let's put this all to use.

Write the following code within an HTML file:

```
<!DOCTYPE html>
<html>
    <body>
        <p id="Keywords_and_Constructors"></p>
        <script>
        function Vehicle(Make, Model, Year, Color) {
            this.Vehicle_Make = Make;
            this.Vehicle_Model = Model;
            this.Vehicle_Year = Year;
            this.Vehicle_Color = Color;
        }
        var Jack = new Vehicle("Dodge", "Viper", 2021, "Red");
        var Emily = new Vehicle("Jeep", "Trail Hawk", 2020, "White and Black");
        var Erik = new Vehicle("Ford", "Pinto", 1971, "Mustard");
        document.getElementById("Keywords_and_Constructors").innerHTML =
        "Erik drives a " + Erik.Vehicle_Color + "-colored " + Erik.Vehicle_Model
        + " manufactured in " + Erik.Vehicle_Year;
        </script>
    </body>
</html>
```

Save and execute your code. This should print the following string:

```
Erik drives a Mustard-colored Pinto manufactured in 1971
```

Let's inspect our code. The function "Vehicle()" is an object constructor.

RESERVED WORDS

In JavaScript, there are certain words you can't use as variables, labels (names assigned to sections of code), or functions. Examples of reserved words are: "true" and "false."

If you're interested, here is a link to a list of the reserved words in JavaScript:

w3schools.com/js/js_reserved.asp

The reason you cannot use these words is that they already mean something else – they are reserved for JavaScript.

END OF CHAPTER CHALLENGE

Write a program that includes the following:

- A class,
- A constructor,
- The "new" keyword, and
- The "this" keyword.

CHAPTER TWENTY FIVE
IDENTIFIERS AND LITERALS

An "identifier" is a name for something. In JavaScript, identifiers are the names of variables, functions, keywords, and labels. In the following code, "X" is the identifier:

```
var X = 10
```

In contrast to an identifier, a literal is something that represents a value within source code. Source code is the version of a computer program as it is originally written (i.e., typed into a computer in some programming language) by the designer of the program. Normally, when the programmer is done making the program, she has the computer convert the code she wrote the program in (the "source code") to another format that is easier and faster for the computer to use. In that new format, the program can't be easily understood by people – but it is very useful to the computer. It also can't be modified by other programmers at that point.

Look at this code:

```
var X = 10
var Y = "Charlie"
```

The 10 is the integer (whole number) literal, Charlie is the string literal, and "X" and "Y" are the identifiers.

A literal is the data exactly as it is meant to be processed. Whereas an identifier is a name, the literal is the value itself.

NESTED FUNCTIONS

Nested refers to something contained within something else. This can be a program within a program or a set of instructions inside another set of instructions.

Sometimes the subprogram needs some information from the main program in order to perform its tasks. When the subprogram is created, its description might include this information. That information is called the "parameters" of the subprogram.

The parameters of the nested entity are explained within the larger (and enveloping) entity. It's nested because it is placed inside something else.

For example, a nested procedure would be a set of commands that is contained within a larger procedure.

In JavaScript, functions have access to the functions that are above them in the code. A nested function is a function under another function that is connected somehow.

To see this in action, write this code:

```
<!DOCTYPE html>
<html>
    <body>
        <p id="Counting"></p>
        <script>
        document.getElementById("Counting").innerHTML = Count();
        function Count() {
            var Starting_point = 9;
            function Plus_one() {Starting_point += 1;}
            Plus_one();
            return Starting_point;
        }
        </script>
    </body>
</html>
```

Save and execute this code. "10" should be displayed. In our code, the Plus_one() function was the nested function.

<u>SCOPE</u>

In programming, code can have "scope." Simply put, the scope is the functions, variables, and objects you have access to. Scope can be limited. Variables have scope in that they can either be accessed by one, more than one, or all functions in a program. The scope of variables is either "local" or "global."

In JavaScript, a global variable can be accessed from any function within the program, whereas a local variable is only accessed by the function it is assigned to.

Global variables are declared outside of functions, and local variables are declared inside of functions.

A global variable would be written as follows (write this code):

```
<!DOCTYPE html>
<html>
    <body>
        <script>
        var X = 10;
        function Add_numbers_1() {
            document.write(20 + X + "<br>");
        }
        function Add_numbers_2() {
            document.write(X + 100);
        }
        Add_numbers_1();
        Add_numbers_2();
        </script>

    </body>
</html>
```

Save and execute your code. It should return "30" and "110." The variable X was assigned the value 10 outside of our function, but we still accessed it. Therefore, the above is an example of a global variable. As a reminder, <br> is the break tag and creates a line break in text (like pressing enter on your keyboard).

Now, let's try a local variable. Write this code:

```
<!DOCTYPE html>
<html>
    <body>
        <script>
        function Add_numbers_1() {
            var X = 10;
            document.write(20 + X + "<br>");
        }
        function Add_numbers_2() {
            document.write(X + 100);
            }
        Add_numbers_1();
        Add_numbers_2();
        </script>
    </body>
</html>
```

Save and execute your code. This time, it only returns "30" because the variable was local – meaning it was written within the function Add_numbers_1 and couldn't be accessed outside of it.

Let's say you wrote the above code and didn't understand why Add_numbers_2 didn't display a result. We could use the console.log() to help us debug our code.

Change your code to the following:

```
<!DOCTYPE html>
<html>
    <body>
        <script>
        function Add_numbers_1() {
            var X = 10;
            console.log(15 + X);
        }
        function Add_numbers_2() {
            console.log(X + 100);
        }
        Add_numbers_1();
        Add_numbers_2();
        </script>
    </body>
</html>
```

Save and execute your code. You'll notice that no result is shown. Open the console, and there it is! It gives you the error "X is not defined."

END OF CHAPTER CHALLENGE

Perform the following actions:

- Write your own nested function,
- Assign a local variable, and
- Assign a global variable.

"If statements" are a type of conditional statement that specifies that a section of code is to be executed *if* a condition is true.

Write this code:

```
<!DOCTYPE HTML>
<html>
    <body>
        <script>
        if (1 < 2) {
            document.write("The left number is smaller than the right number.")
        }
        </script>
    </body>
</html>
```

Save and execute your code. Since 1 is smaller than 2, the string is displayed.

The `Date().getHours()` method returns the hour in the specified date according to the local time, and the hours are listed as integers between 0 and 23. For example: 18 is equal to 6:00 p.m., 12 is noon, etc.

We can get creative with "if" statements. Write this code:

```
<!DOCTYPE html>
<html>
    <body>
        <p id="Greeting">How are you this evening?</p>
        <script>
        if (new Date().getHours() < 18) {
            document.getElementById("Greeting").innerHTML = "How are you today?";
        }
        </script>
    </body>
</html>
```

Save and execute your code. In this program, we said, "If it is later than (greater than) 6:00 p.m. when I ran my code, display 'How are you this evening?' If it is earlier than 6:00 p.m. when I ran my code, 'How are you today?' will display."

So, what if we want more choices?

ELSE STATEMENTS

The else statement specifies a block of code that will be executed if the condition is false (opposite of the if statement).

Write this code:

```
<!DOCTYPE html>
<html>
    <body>
        <p>Write your age:</p>
        <input id="Age" value="" />
        <p id="How_old_are_you?"></p>
        <button onclick="Age_Function()">Click here</button>
        <script>
        function Age_Function() {
            Age = document.getElementById("Age").value;
            if (Age >= 18) {
                Vote = "You are old enough to vote!";
            }
            else {
                Vote = "You are not old enough to vote!";
            }
            document.getElementById("How_old_are_you?").innerHTML = Vote;
            }
        </script>
    </body>
</html>
```

Save and execute your code. You just created a program that determines if a person can vote!

Now, what if we want even more choices?

ELSE IF STATEMENTS

The "else if" statement follows an "if" statement and is executed if the "if" statement is found to be false. For example:

- If hungry, eat.
- Else if thirsty, drink.
- Else, rest.

Write this code:

```html
<!DOCTYPE html>
<html>
    <body>
        <p id="Time_of_day"></p>
        <script>
        function Time_function() {
            var Time = new Date().getHours();
            var Reply;
            if (Time < 12 == Time > 0) {
                Reply = "It is morning time!";
            }
            else if (Time > 12 == Time < 18) {
                Reply = "It is the afternoon.";
            }
            else {
                Reply = "It is evening time.";
            }
            document.getElementById("Time_of_day").innerHTML = Reply;
        }
        Time_function();
        </script>
    </body>
</html>
```

Save and run your code. You made a program that pulls the time from your computer and tells you what time of day it is!

END OF CHAPTER CHALLENGE

Perform the following:

- Write your own function that contains an "if" statement, "else if" statement and "else" statement.

CHAPTER TWENTY SEVEN
STRING METHODS

JavaScript has string methods that allow you to perform tasks with strings.

As covered earlier, programming languages have certain data types built into them. These built-in data types are the most basic data types in the programming language and are referred to as primitive data types. One of the key traits of a primitive data type is that it cannot be simplified any further. A primitive data type is a predefined type of data that is built into the language.

There are five primitive data types in JavaScript – these are:
1) String,
2) Number,
3) Boolean,
4) Null, and
5) Undefined.

Everything else (non-primitive data types) are objects. Objects are created by the programmer, not predefined and built into the programming language.

CONCAT() METHOD

The concat() method concatenates (connects) two or more strings. Write this code:

```
<!DOCTYPE html>
<html>
    <body>
        <p id="Concatenate"></p>
        <script>
        var part_1 = "I have ";
        var part_2 = "made this ";
        var part_3 = "into a complete ";
        var part_4 = "sentence.";
        var whole_sentence = part_1.concat(part_2, part_3, part_4);
        document.getElementById("Concatenate").innerHTML = whole_sentence;
        </script>
    </body>
</html>
```

Save and execute your code. It displays the sentence as a whole!

SLICE() METHOD

The slice() method is a string method that extracts a section of a string and then returns the extracted section in a new string.

Write this code:

```
<!DOCTYPE html>
<html>
    <body>
        <p id="Slice"></p>
        <script>
        var Sentence = "All work and no play makes Johnny a dull boy.";
        var Section = Sentence.slice(27,33);
        document.getElementById("Slice").innerHTML = Section;
        </script>
    </body>
</html>
```

Save and execute your code. The numbers choose which characters in your string will be cut out and displayed. Your code displays "Johnny" because those are the characters that are located between 27-33.

Here are two points to keep in mind:
1) Computers start counting at 0 – not 1.
2) The spaces are included in the count.

END OF CHAPTER CHALLENGE

You will now be assigned some tasks we haven't taught yet – see if you can find the solutions with a bit of online research:

1. Utilize the toUpperCase() method, and
2. Utilize the search() method.

CHAPTER TWENTY EIGHT
NUMBER METHODS

Number methods assist you in working with numbers in JavaScript. There are several number methods. One of these is the `toPrecision()` method which returns a string as a number of a specified length. Write this code:

```
<!DOCTYPE html>
<html>
    <body>
        <p id="Precision"></p>
        <script>
        var X = 12938.3012987376112;
        document.getElementById("Precision").innerHTML =
            X.toPrecision(10);
        </script>
    </body>
</html>
```

Save and run your code. You should get "12938.30130." There are many number methods – the above two should give you an idea of how these work.

END OF CHAPTER CHALLENGE

You will now be assigned some tasks we haven't taught yet – research online for solutions:

```
1. Utilize the toFixed() method.
2. Utilize the valueOf() method.
```

Here is a helpful link: w3schools.com/js/js_number_methods.asp

CHAPTER TWENTY NINE
FINAL CHALLENGE

To conclude this book, we're going to put it all together by creating a timer that will countdown by seconds. Here we go!

First, replace any existing code in your editor with the following:

```
<!DOCTYPE html>
<html>
    <body>
        <p>How many seconds would you like to set your alarm for?</p>
        <input id="seconds" value="" />
        <button onclick="countdown()">Click here</button>
        <p id="timer"></p>
    </body>
</html>
```

The idea here is that the user will be asked how many seconds he/she would like to set the timer for. There is an input box available where a value may be entered. A button is also created so that when clicked, it will send the value out to be processed. The result will be a displayed countdown timer.

Remember that the input tag deals with information that the user enters. The id "minutes" and the value "" are associated with the JavaScript code that we'll see shortly.

The "button" tag will create a button that, when clicked, will look in the JavaScript code for the "countdown()" function. The result that the countdown function produces will depend upon what the user submitted when he/she clicked the button we made.

The tag with the id "timer" wants to know the result of the countdown function so that it can display a timer for the user to see. It will retrieve this information from the associated JavaScript.

So, let's take a look at the HTML code again, but this time, with the addition of the JavaScript code:

```html
<!DOCTYPE html>
<html>
    <body>
        <p>How many seconds would you like to set your alarm for?</p>
        <input id="seconds" value="" />
        <button onclick="countdown()">Click here</button>
        <p id="timer"></p>
        <script>
        function countdown() {
            var seconds = document.getElementById("seconds").value;

            function tick() {
                seconds = seconds - 1;
                timer.innerHTML = seconds;
                setTimeout(tick, 1000);
            if(seconds == -1){
                alert("Time's up!");
            }
                }
            tick();
        }
        </script>
    </body>
</html>
```

You'll notice right away that there are two separate functions in our JavaScript code. In fact, the "tick" function is sitting inside of the "countdown" function! As a reminder, this is called "nesting functions."

The countdown function is called just one time, and it does only two things:

1. It sets a variable named "seconds" to the value that the user inputs.
2. It creates and runs the tick function.

The tick function, on the other hand, has several steps to execute and will end up repeating multiple times.

Immediately after the countdown function has received the input value from the user, the tick function begins laying out its instructions. It's important to recognize that the tick function is not actually being executed yet. It is simply letting its instructions be made known in case the function ever actually needs to be used. Once the interpreter reads the tick function, it steps out and sees the "tick()" instruction. Upon seeing this function, the Interpreter knows to immediately go back through the tick function. However, this time, instead of simply reading the function, it is actually going to run the function.

Take a look at the variable "seconds" within the tick function. It is taking the value "seconds" from the countdown function and subtracting 1 from it. The reason that this value needs to be subtracted by 1 is that each time this function runs, it will last one second (as we will see), and we need the variable "seconds" to subtract 1 each time a second goes by.

Right after 1 second is subtracted from the user's input the first time through the tick function, that same number that has been subtracted by 1 will now be displayed for the user to see. This happens because the `timer.innerHTML` is being assigned this number. Remember how we put "id=timer" in one of our HTML tags? That tag now holds this number that we assigned it.

The statement "`setTimeout(tick, 1000);`" simply means that the program is pausing for 1,000 milliseconds (i.e., 1 second). This is important because this function is going to potentially repeat multiple times – and every time it repeats, it needs to do so for a duration of exactly one second.

We now have our conditional statement. What it's telling us is that if the user's input ever counts all the way down to -1, stop everything and run the alert box. You may be asking yourself, *Why -1 instead of 0?* The reason is that when the interpreter sees -1, it hasn't actually displayed it to the user yet. The user is still seeing a 0. Therefore, before the -1 is displayed, the alert box will show up, preventing any other code from running. As a note, if the user clicks the OK button on the alert box, he/she will see the timer continue to count down into the negatives.

Now, if the user's input does not equal -1, the interpreter will simply step out of the function like before. But wait; it's going to run into that "tick()" statement again! Exactly. Do you see how our function will continue to repeat now? That is, until our conditional statement is met – wherein the alert box is displayed.

Try it out for yourself and don't worry if you have to read the code through several times before it really begins to sink in!

FINAL CHALLENGE

Here is your final assignment:

```
Through online research, find out how to end the program (i.e, stop the
countdown) when a user clicks the OK button on the alert box.
```

CHAPTER THIRTY
CONGRATULATIONS!

You have completed our book! Well done on your hard work and persistence!

You should now have a basic understanding of JavaScript and coding.

As the next step, we recommend enrolling in a Tech Academy coding boot camp. Our coding boot camps were designed like this book: for beginners and assuming no prior knowledge or experience.

Here are some of the reasons thousands of students have chosen to enroll at The Tech Academy:

1) Our curriculum – it's modern, robust (strong; holds up over time), and covers in-demand technologies. Our programs are thorough and cover more than just 1-2 languages. Our comprehensive curriculum ensures that students aren't pigeon-holed (restricted to an exclusive category) within a small skillset. Understanding a large array of technologies not only prepares graduates for the workforce, it assists them greatly in picking up new tech skills in the future.

2) We price our boot camps affordably. In fact, our tuition is less than the national average weekly cost of coding boot camps.

3) We require no technical background or experience. You don't have to already know coding to learn to code. As long as students can read, write and perform basic math, they can succeed at The Tech Academy.

4) We have a stellar online presence. Our average review rating across the top review sites ranges from 4.5-4.9 stars. We have received the Best Coding Boot Camp award several years in a row from SwitchUp.Org and CourseReport.Com (the top two boot camp review sites) and are included on several other top coding boot camp lists as well. We were also chosen as "The World's Best Code School" by the television show "World's Greatest." All of these awards and reviews are based on feedback from students and graduates of our programs.

5) We are extremely flexible. Students choose their own study schedule. They can study from home, at one of our campuses or both. They have online access to their program 24 hours a day. This ensures that students can enroll regardless of their life circumstances. The fact that the programs are self-paced is an aspect of our flexibility – students can blast through content they know well already, and take their time with new concepts. An additional factor in our flexibility is that we offer open enrollment, which means students can start anytime.

6) Our admissions staff and enrollment process are transparent and helpful. We answer questions and are polite, giving students a positive experience from the start.

To get started, visit The Tech Academy's website: learncodinganywhere.com

INDEX

INDEX

OTHER READING

Be sure to check out other Tech Academy books, which are all available for purchase on Amazon:

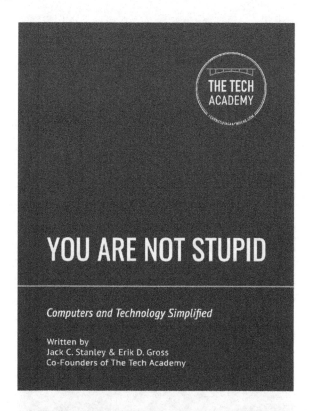

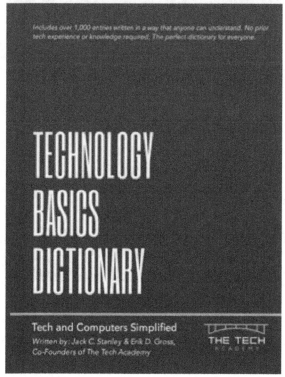

LEARN CODING BASICS IN HOURS WITH SMALL BASIC

Programming for Absolute Beginners

Written by: Jack C. Stanley & Erik D. Gross,
Co-Founders of The Tech Academy

LEARN CODING BASICS IN HOURS WITH PYTHON

Programming for Absolute Beginners

Written by: Jack C. Stanley & Erik D. Gross,
Co-Founders of The Tech Academy

THE TECH
ACADEMY

PROJECT MANAGEMENT
HANDBOOK

Simplified Agile, Scrum and DevOps for Beginners

Written by
Jack C. Stanley & Erik D. Gross
Co-Founders of The Tech Academy

Made in the USA
Monee, IL
27 September 2020